GeoJournal Library

Volume 124

More information about this series at http://www.springer.com/series/6007

Enrico Gottero
Editor

Agrourbanism

Tools for Governance and Planning of Agrarian Landscape

Springer

Editor
Enrico Gottero
IRES Piemonte
Turin, Italy

ISSN 0924-5499 ISSN 2215-0072 (electronic)
GeoJournal Library
ISBN 978-3-030-07062-5 ISBN 978-3-319-95576-6 (eBook)
https://doi.org/10.1007/978-3-319-95576-6

This Springer imprint is published by the registered company Springer Nature Switzerland AG
The registered company address is: Gewerbestrasse 11, 6330 Cham, Switzerland

Contents

Editor and Contributors

About the Editor

Enrico Gottero (Editor) is lecturer in Landscape and Regional Planning at the Interuniversity Department of Regional and Urban Studies and Planning (Politecnico di Torino), as well as researcher at the Institute of Economic and Social Research of Piedmont (IRES). His main interests focus on the interaction between agriculture and rural-agrarian landscapes, urban-rural linkages, farmland preservation, urban agriculture, agri-food policies and practices. He is also author of several publications on these topics, including at international level.

Contributors

Carlo Alberto Barbieri DIST, Interuniversity Department of Regional and Urban Studies and Planning, Politecnico and Università Di Torino, Turin, Italy

Christopher Bryant School of Environmental Design and Rural Development, University of Guelph, Guelph, ON, Canada

Wayne Caldwell School of Environmental Design and Rural Development, University of Guelph, Guelph, ON, Canada

Claudia Cassatella DIST, Interuniversity Department of Regional and Urban Studies and Planning, Politecnico and Università di Torino, Turin, Italy

Amanda Jennifer Chandra Department of Architecture, National University of Singapore, Singapore, Singapore

Jessica Ann Diehl Department of Architecture, National University of Singapore, Singapore, Singapore

Sara Epp School of Environmental Design and Rural Development, University of Guelph, Guelph, ON, Canada

David Fanfani DIDA—Architecture Department, Florence University, Florence, Italy

Carolina Giaimo DIST, Interuniversity Department of Regional and Urban Studies and Planning, Politecnico and Università Di Torino, Turin, Italy

Enrico Gottero IRES Piemonte, Torino, Italy

Nobuko Kawaguchi Graduate School of Environmental Studies, Nagoya University, Nagoya, Japan

Frank Lohrberg Department of Landscape Architecture Aachen, RWTH Aachen University, Aachen, Germany

Marco Marchetti Dipartimento di Bioscienze e Territorio, Università degli Studi del Molise, Pesche, Italy

Susan Parham Environment and Agriculture Group, UH Geography, University of Hertfordshire, Hatfield, England

Giacomo Pettenati Dipartimento di Culture, Politica, Società, Università di Torino, Turin, Italy

Carlo Rega ITERAS—Research Centre for Sustainability and Territorial Innovation, Bari, BA, Italy; European Commission—Joint Research Centre, Directorate D—Sustainable Resources, Ispra, VA, Italy

Stefano Salata DIST, Interuniversity Department of Regional and Urban Studies and Planning, Politecnico and Università Di Torino, Turin, Italy

Lorenzo Sallustio Dipartimento di Bioscienze e Territorio, Centro di Ricerca per le Aree Interne e gli Appennini (ArIA), Università degli Studi del Molise, Campobasso, Italy; CREA Research Centre for Forestry and Wood, Arezzo, Italy

Marco Santangelo DIST, Interuniversity Department of Regional and Urban Studies and Planning, Politecnico and Università di Torino, Turin, Italy

Bianca Maria Seardo DIST, Interuniversity Department of Regional and Urban Studies and Planning, Politecnico and Università di Torino, Turin, Italy

Hiroyuki Shimizu Graduate School of Environmental Studies, Nagoya University, Nagoya, Japan

Ching Sian Sia Department of Architecture, National University of Singapore, Singapore, Singapore

Chika Takatori Graduate School of Environmental Studies, Nagoya University, Nagoya, Japan

Matteo Vizzarri Dipartimento di Bioscienze e Territorio, Università degli Studi del Molise, Pesche, Italy

Chapter 1
Approaching a Vision of Agrarian Urbanism: Innovative Domains, Key Definitions and Concepts

Enrico Gottero

This book is the outcome of several years of research in the interwoven fields of agricultural urbanism and landscape studies. The innovative topic of research is at the heart of an intense and controversial international debate involving several disciplines, such as urbanism, agricultural science, landscape planning/design and geography. It focuses on the interaction between topics, such as agriculture (agricultural and food policies), spatial planning, environmental assessment, urban-rural linkages and farmland preservation, besides urban and peri-urban agricultural practices. Some of these recent studies, but also each scientific exchange or debate with the authors of this book, have defined a firmly grounded preliminary foundation for the editorial project based on the agroecosystem's heterogeneity, variety and complexity. Hence, the theoretical and methodological approach to this interrelated research field establishes a starting point, an interesting study opportunity that, however, still requires an experimental phase and much time to be processed by the scientific community.

Studies lately conducted on these issues by many scholars reconsider agriculture with a planning-based approach, after several years of indifference, and mainly stem from recent dramatic events, such as the economic crisis in many European countries. Moreover, austerity policies and neoliberal models have given rise, over the last decade, to several socio-spatial and socio-ecological implications, and to effects on individual and collective human wellbeing, both in urban and rural areas (Knieling and Othengrafen 2016). At the same time, capitalist urbanisation and the development of the agro-industry (intensification and industrialisation) have expanded the gap and dichotomy between humans and nature, and also between urban and rural context. The rupture between town and countryside, as well as the rift between agriculture and other forms of land use, often caused by important public and private interests that compete with each other for access and use of the

E. Gottero (✉)
IRES Piemonte, Via Nizza 18, 10100 Torino, Italy
e-mail: e.gottero@ires.piemonte.it

E. Gottero (ed.), *Agrourbanism*, GeoJournal Library 124,
https://doi.org/10.1007/978-3-319-95576-6_1

same land base (Connell 2009), have given rise to an urban and rural system of asymmetric and unequal socio-economic and environmental relations (Dehaene et al. 2016). Finally, considering the latest estimates of the world population—the FAO[1] claims that the number of people living on our planet will rise remarkably by 2050 (from 6.3 to 9 billion)—management and use of resources (food, water, energy, etc.) certainly cannot refer to the model of life adopted to date in Western countries. Studying the production and consumption of food, energy and irreproducible resources, as well as public health issues (obesity, heart disease, physical and mental well-being) has, therefore, become urgent for mankind. Hence the impelling and crucial need to define new development options based on new sustainable solutions, especially in the Global North, and in the context of great political and social uncertainty (Boukharaeva and Marloie 2006).

However, spatial planning models recently adopted to overcome these issues have often indicated conflicting results. On the one hand the functionalist model, the socio-urbanist theory of community and neighbourhood unity, as well as the "city in the countryside" framework have underscored unrealistic solutions and several limitations (Matos and Batista 2013). Other theories have, instead, been successful such as, for example, the green belt concept or theories about the "garden city" of Howard originated in England in the late nineteenth century in response to significant socio-economic changes. They were not only based on the productive function of agriculture but, especially, on its meaning in terms of green spaces around the city. Several experiments have been conducted in this regard, including the "Ceinture verte" of Paris (Vidal and Fleury 2009), the GrunGurtel of Frankfurt and Munich, the "Anella verde" of Barcelona, and the Finger Plan of Copenhagen (see also Caspersen and Olafsson 2010; Vejre 2008; Zasada et al. 2011). Other scholars have recently studied this topic in various ways. Some approaches focus on food sovereignty and provision (Dansero and Pettenati 2015; Gorgolewski et al. 2011; Ilieva 2016; Morgan and Sonnino 2010; Parham 2015; Wascher et al. 2015), on social equity (Meyer-Renschhausen 2014; Pinto et al. 2010), on the promotion of a continuous productive urban landscape (Viljoen and Bohn 2014), on urban agriculture (Mougeot 2005; Mininni 2017; Lohrberg et al. 2016), on the "back to earth"[2] theory, on building a resilience concept and transition movement (Barthel et al. 2015; Clausen 2015; Sage 2014), on the permaculture theory (Whitefield 2011), on the model of biophilic cities (Beatley 2010) or, specifically, on the interaction between landscape/spatial planning and agriculture (De La Salle and Holland 2010; Duany 2012; Long 2015; Poli 2013; Rega 2014; Waldheim 2006, 2016; Wise 2013).

[1]See: http://www.fao.org/statistics/en/ (Accessed: 12/04/2016).

[2]In this regard see also the monographic volume of the Italian review "Scienze del Territorio" on the issue "Back to Earth" (1/2013).

However, to date, we get the feeling that we are still in an empirical phase of experimental research for new tools, not only to contain the expansion of the city, but also to restore sustainable food, social and cultural balance between town and country.

Understanding complex relations, filling up the lack of codified tools and of a holistic and unitary vision (multidisciplinary and interdisciplinary), as well as responding to new emerging social demands for green spaces, food quality, cultural services and lifestyles, especially in many European countries, will establish the foundations for a new multidisciplinary approach. The latest agricultural planning experiences, such as, for example, the metropolitan region of Vancouver (Boucher 2009), the first multidisciplinary examples of the agri-urban programmes of Ile-de-France (Vidal 2009), and the academic activity of the Versailles Architecture School belong to this context. On the one hand, these cases provide evidence of the attempt to fill an ideological gap by integrating functional urban planning and agriculture, or all activities related to agricultural production and urban uses, especially in order to underscore mutual benefits in terms of ecosystem and socio-economic services (Boucher 2009). On the other hand, they reveal the need for new transdisciplinary and specific skills and for new professional figures that do not explicitly fall into the categories of either urbanism, agriculture, landscape or geography but, instead, constitute an interconnection between these various disciplines. In others words, the concept of agrarian urbanism (especially in this book) is based on a new urban-rural balance, on the reconciliation of "rural-urban rift", on a new partnership between emerging forms of agriculture and new multidisciplinary approaches, a paradigm that can be interpreted as a new form of city but also as renewed rurality. The concept of *agrourbanism*, also called *agriurbanisme* (Bronzini and Bedini 2015; Vidal and Fleury 2009) or *agricultural urbanism*, or *urbanisme agricole* (Boucher 2009) is based on a multiform and multifunctional economy made up of local products and high environmental quality, on a low growth model as opposed to the industrial one (Donadieu 2005; Fanizza 2015; Ferraresi 2013; Ingersoll 2012). In these terms, an agrarian urbanism approach could offer the opportunity to rethink unsustainable models and to improve environmental conditions, as well as a chance to improve human health, quality of life and wellbeing both in cities and in the countryside. It outlines not only the needs of cities but also those of agriculture without distinction, avoiding social, economic and cultural inequalities (Vidal 2009). It could also be an opportunity to ensure basic agro-ecosystem functions (Parham 2015; Torquati and Giacchè 2010).

Thus, adopting the Agrarian Urbanism approach as a tool for alternative and sustainable planning is a viable path that few scholars have attempted to outline. However, especially in Europe, there are no specific tools to check this hypothesis and to address a planning system today. There are no consistent and integrated approaches to the mending process of the urban-rural rift. Starting from this deficiency, cities and regions need to fill this gap through a new model of governance and planning designed to overcome socio-economic crises, face new societal challenges and respond to the new demands of countryside, recreational venues, landscape and quality food products by adopting a holistic approach. This book, therefore, explores the various forms of agricultural and urban planning, as well as the main dimensions of innovation associated with this approach in terms of governance (policies and practices), spatial dimensions and functional features. At the same time, this book verifies the practicability and transferability of this model to European governance systems, to also ensure basic agroecosystem standards to cities and citizens, such as food, cultural services and biodiversity.

From this perspective, the first section of the book explores the various meanings of the concept of agrarian urbanism in several contexts and scales, also taking into account the international point of view through a multidisciplinary and transcalar approach. In Chap. 2, Epp et al. discuss the problems and potential of farmland preservation in Canada, while Chaps. 3 and 4 explore two experiences of the eastern world. Takatori et al. and Diehl et al. have respectively addressed two topics of great interest for agrarian urbanism applied with a heterogeneous approach, precisely rural landscape management (through a Japanese case study) and food demand and security issues (in Singapore). Section Two investigates the governance background of agrarian urban spaces. Santangelo explains the relationship between urban and rural, while Rega shows how planning and managing agricultural areas could interact with effective implementation of the multifunctional connectivity of Green Infrastructures. Gottero and Cassatella, instead, underscore the different spatial dimensions of European rural development policies, while Parham reveals increasingly significant interactions between food and urbanism in the context of the new food planning system. Section Three explores several forms and places in which the concept of agrarian urbanism manifests itself, precisely in urban and peri-urban spaces, as well as in professional and non-professional forms, as stated by Lohrberg. Fanfani and Pettenati study another two ways of fostering agrarian urban values, namely one centred on agricultural parks, and the other on the new emerging local urban food plans. Finally, in the same section, Seardo describes a possible way of implementing the agrarian urbanism model, especially underscoring the importance of social perception as a tool that can preserve rural and agricultural landscapes. The last part of the book presents new functions and spatial tools to solve the issue of agrarian urbanism with a territorial approach. Marchetti et al. analyse and discuss various methods to reduce soil consumption and evaluate its impact on agricultural productivity, while Giaimo et al. outline new solutions to integrate the assessment of ecosystem services in land use regulation at a municipal level, with particular focus on those associated with the agrarian urban system. Finally, Gottero and Cassatella discuss some applications at different scales by adopting a map-based approach to prove the efficacy and usefulness of certain spatial tools, especially in the framework of supra-local planning. In the last chapter the editor justifies this perspective by discussing certain key issues and the ideas behind this vision, and by exploring some reasons that should underlie this approach.

References

Barthel, S., Parker, J., & Ernstson, H. (2015). Food and green space in cities: A resilience lens on gardens and urban environmental movements. *Urban Studies*, 1321–1338.

Beatley, T. (2010). *Biophilic cities: Integrating nature into urban design and planning*. Washington D.C.: Island Press.

Boucher, I. (2009). D'agriculture urbaine à urbanisme agricole: Une participation au développement durable, une contribution à la production alimentaire, http://www.mamrot.gouv.qc.ca/pub/observatoire_municipal/veille/agriculture_urbaine.pdf. Accessed February 02/2016.

Boukharaeva, L. M., & Marloie, M. (2006). Family urban agriculture as a component of human sustainable development. *Cab Reviews: Perspectives in Agriculture, Veterinary Science, Nutrition and Natural Resources*, 1.

Bronzini, F., & Bedini, M. A. (2015). L'abbraccio città-campagna. *Archivio si Studi Urbani e Regionali, 122,* 60–76.

Caspersen, O. H., & Olafsson, A. S. (2010). Recreational mapping and planning for enlargement of the green wedges in Greater Copenhagen. *Urban Forestry and Urban Greening, 9*(2), 101–112.

Clausen, M. (2015). Urban agriculture between pioneer use and urban land grabbing: The case of "Prinzessinnengarten" Berlin. *Cities and the Environment* (CATE), *8*(2), 1–5 (5).

Connell, D. J. (2009). Planning and its orientation to the future. *International Planning Studies, 14* (1), 85–98.

Dansero, E., & Pettenati, G. (2015). Alternative food networks as spaces for the re-territorialisation of food: The case of Turin. In G. Cinà., & E. Dansero (Eds.), *Localizing urban food strategies: Farming cities and performing rurality. Proceedings of 7th International Aesop Sustainable Food Planning Conference*, , (pp. 552–565) Torino: Politecnico di Torino. October 7–9, 2015.

Dehaene, M., Tornaghi, C., & Sage, C. (2016). Mending the metabolic rift: Placing the "urban" in urban agriculture. In F. Lohrberg, L. Licka, L. Scazzosi, & A. Timpe (Eds.), *Urban Agriculture Europe* (pp. 174–177). Berlin: Jovis Verlag.

De La Salle, J. M., & Holland, M. (2010). *Agricultural urbanism: Handbook for building sustainable food and agriculture systems in 21st century cities*. Winnipeg: Green Frigate Books.

Donadieu, P. (2005). Dall'utopia alla realtà delle campagne urbane. *Urbanistica, 128,* 15–20.

Duany, A. (2012). *Garden cities: Theory and practice of agrarian urbanism*. London: The Prince's Foundation for the Built Environment.

Fanizza, F. (2015). La città giardino secondo l'agrarian urbanism: Multifunzionalità in agricoltura e benessere sociale. *Sociologia Urbana e Rurale, 106,* 137–147.

Ferraresi, G. (2013). Neo-rurality: Roots of future in the field. In *Scienze del Territorio* (pp. 71–78). Firenze: Firenze University Press. (1/2013).

Gorgolewski, M., Komisar, J., & Nasr, J. (2011). *Carrot city: Creating places for urban agriculture*. New York: Monacelli Press.

Ilieva, R. T. (2016). *Urban food planning: Seeds of transition in the global North*. London-New York: Routledge.

Ingersoll, R. (2012). Urban agriculture. *Lotus, 149*(74), 105–117.

Knieling, J., & Othengrafen, F. (Eds.). (2016). *Cities in crisis: Socio-spatial impacts of the economic crisis in Southern European cities*. New York: Routledge.

Lohrberg, F., Licka, L., Scazzosi, L., & Timpe, A. (Eds.). (2016). *Urban agriculture Europe*. Berlin: Jovis Verlag.

Long, C. (2015). Agricultural urbanism toolkit, extension and outreach publications. Book 12, on: http://lib.dr.iastate.edu/extension_pubs/12. Accessed December 29, 2016.

Matos, R. S., & Batista, D.S. (2013). Urban agriculture: The allotment gardens as structures of urban sustainability. In M. Özyavuz (Ed.), *Advances in landscape architecture*. InTech, pp. 457–512, http://www.intechopen.com/books/advances-in-landscape-architecture/urban-agriculture-the-allotment-gardens-as-structures-of-urban-sustainability. Accessed January 20, 2016.

Meyer-Renschhausen, E. (2014). Community gardening in Berlin and New York: A new eco-social movement. In A. Viljoen & K. Bohn (Eds.), *Second nature urban agriculture: designing productive cities* (pp. 146–153). London & New York: Routledge.

Mininni, M. (2017). *Matera Lucania*. Quodlibet: Laboratorio città paesaggio. Macerata.

Morgan, K., & Sonnino, R. (2010). The urban foodscape: World cities and the new food equation. *Cambridge Journal of Regions, Economy and Society, 3*(2), 209–224.

Mougeot, L. J. A. (Ed.) (2005). *Agropolis: The social, political, and environmental dimensions of urban agriculture*, Ottawa: Earthscan and the International Development Research Centre (IDRC).

Parham, S. (2015). *Food and urbanism: The convivial city and a sustainable future*. London: Bloomsbury.

Pinto, B., Pasqualotto, A., & Levidow, L. (2010). Community supported urban agriculture: The orti solidali project in Rome. *Urban Agriculture Magazine, 24*, 58–60.

Poli, D. (Ed.). (2013). *Agricoltura paesaggistica. Visioni, metodi, esperienze*. Firenze: Firenze University Press.

Rega, C. (Ed.). (2014). *Landscape Planning and rural development: Key issues and options towards integration*. Cham Heidelberg New York Dordrecht London: Springer.

Sage, C. (2014). The transition movement and food sovereignty. *Journal of consumer culture, 14* (2), 254–275.

Torquati, B., & Giacchè, G. (2010). Rapporto città—campagna e sviluppo rurale. *Agriregionieuropa, anno, 6*(20), 6–9.

Vejre, H. (2008). *Models for safeguarding urban fringe open landscapes—the balance between public and private. Case study of management models for urban green space around Copenhagen*. Conference Proceedings "Rurality near the city", Leuven.

Vidal, R. (2009). Entre ville et campagne, un paysage à inventer. Coll. "Ateliers techniques du paysage", Conseil Régional de Bre-tagne, Rennes.

Vidal, R., & Fleury, A. (2009). Aménager les relations entre la ville et l'agriculture, de nouveaux enjeux territoriaux et une nouvelle approche «agriurbaniste». *Revue Urbia, n. 8*, 127–142.

Viljoen, A., & Bohn, K. (Eds.). (2014). *Second nature urban agriculture: Designing productive cities*. London & New York: Routledge.

Waldheim, C. (2006). *The landscape urbanism reader*. New York: Princeton architectural press.

Waldheim, C. (2016). *Landscape as urbanism: A general theory*. Oxford: Princeton University Press.

Wascher, D., Kneafsey, M., Pintar, M., & Piorr, A. (2015). *Food planning and innovation for sustainable metropolitan regions*. Synthesis Report, Wageningen: Wageningen University Research.

Whitefield, P. (2011). *The earth care manual: A permaculture handbook for Britain and other temperate climates*. Hampshire: Permanente pubblications.

Wise, E. (2013). A gradual reawakening: Broadacre city and a new american agrarianism. *Berkeley Planning Journal, 26*, 133–149. https://escholarship.org/uc/item/767512g6. Accessed December 29, 2016.

Zasada, I., Fertner, C., Piorr, A., & Nielsen, T. S. (2011). Peri-urbanisation and multifunctional adaptation of agriculture around Copenhagen. *Danish Journal of Geography, 111*(1), 59–72.

Enrico Gottero (Editor) is lecturer in Landscape and Regional Planning at the Interuniversity Department of Regional and Urban Studies and Planning (Politecnico di Torino), as well as researcher at the Institute of Economic and Social Research of Piedmont (IRES). His main interests focus on the interaction between agriculture and rural-agrarian landscapes, urban-rural linkages, farmland preservation, urban agriculture, agri-food policies and practices. He is also author of several publications on these topics, including at international level.

Part I
Agrourbanism from a Global Perspective

Chapter 2 Farmland Preservation and Rural Development in Canada

Sara Epp, Wayne Caldwell and Christopher Bryant

2.1 Introduction

Over the past seventy years, agriculture has been significantly transformed as advances in science and technology, in particular machinery and pesticides, have increased yields and demands on the landscape. Agricultural restructuring is evident throughout North America, as is exemplified by the presence of larger-scale farms influenced by economies of scale. In general, smaller farms have been forced out of agriculture in favour of larger enterprises, as the size of farms has increased, while the number of farms has drastically decreased. In Canada, between 1996 and 2016, while the average size of a farm increased and the amount of land being farmed remained stable, the number of farms decreased by 30%, resulting in over 83,000 fewer farms (Ontario Ministry of Agriculture, Food and Rural Affairs 2017). Table 2.1 provides detailed farm consolidation by province, and it is evident that as the number of farms decreased average farm size has increased.

Across Canada, agriculture and the agri-food industry contribute more than $110 billion to the country's gross domestic product and globally, Canada is the fifth largest exporter of agricultural products (Agriculture and Agri-Food Canada 2017). The economic benefits of the agricultural sector cannot be understated and a stable land base is necessary to sustain agricultural production. Given the vast geographic area of Canada, it is often assumed that much of the nation is comprised of farmland and that this land is abundantly available. In fact, only 7.2% of the country is used for agriculture, as poor soil quality and a northern climate make much of the country unsuitable for agriculture (World Bank 2014). While 7.2% of the country still represents a sizeable amount of farmland, soil quality varies significantly, as does the productive potential of this land. The Canada Land Inventory has mapped

S. Epp (✉) · W. Caldwell · C. Bryant
School of Environmental Design and Rural Development,
University of Guelph, Guelph, ON, Canada
e-mail: sepp@uoguelph.ca

E. Gottero (ed.), *Agrourbanism*, GeoJournal Library 124,
https://doi.org/10.1007/978-3-319-95576-6_2

Table 2.1 Total number of farms and average farm size, by province (Statistics Canada 2017)

Province	Total number of farms		Average farm size (acres)	
	2011	2016	2011	2016
Newfoundland and Labrador	510	407	152	174
Prince Edward Island	1,495	1,353	398	425
Nova Scotia	3,905	3,478	261	263
New Brunswick	2,611	2,255	359	370
Quebec	29,437	28,919	280	280
Ontario	51,950	49,600	244	249
Manitoba	15,877	14,791	1,135	1,192
Saskatchewan	36,952	34,523	1,668	1,784
Alberta	43,234	40,638	1,168	1,237
British Columbia	19,759	17,528	327	365

soil quality, creating seven distinct classes of agricultural land based on the degree of limitation imposed by the soil. Of these classes, Class 1, 2 and 3 soils are considered prime agricultural land, given their high potential for agricultural production, with minimal limitations. In contrast, Class 5 and 6 soils are deemed to have significant limitations for agriculture and Class 7 soils have no agricultural capability. While 5% of Canada is comprised of prime agricultural land, only 0.5% of the county is identified as being Class 1 soil (Walton 2003). A summary of the proportion of Class 1, 2 and 3 soil by province is provided in Table 2.2. As demonstrated by this table, Ontario contains more than 56% of the country's Class 1 soil, with the majority located in the southern portion of the province. This area is

Table 2.2 Amount of Class 1, 2 and 3 soils by province (Connell et al. 2016)

Province	Class 1 (km^2)	Class 2 (km^2)	Class 3 (km^2)	Total (km^2)
Newfoundland and Labrador	–	–	67	67
Prince Edward Island	–	2,626	1,422	4,048
Nova Scotia	–	1,700	10,219	11,919
New Brunswick	–	2,056	13,823	15,879
Quebec	223	10,713	13,625	24,561
Ontario	27,635	23,335	25,567	76,537
Manitoba	2,111	29,617	24,499	56,227
Saskatchewan	12,282	73,341	104,482	190,105
Alberta	6,719	38,701	61,039	106,459
British Columbia	78	1,574	5,270	6,922
Canada	49,048	183,663	260,013	492,724

the most rapidly urbanizing area of the country and much of this Class 1 soil is threatened by development pressure.

Given the pressures of urban development and limited availability of farmland, it is important to understand strategies related to farmland preservation in Canada. Farmland preservation in Canada reflects a diverse variety of strategies ranging from the regulatory to community based approaches that build economic capacity on the farm. While the federal government has a minimal role within planning, the provincial government sets the policies but these are largely interpreted and enforced at the municipal level. Furthermore, agricultural land in Canada is largely privately held and discussions connected to property rights generally do not restrict the ability of the province or municipality to construct policies aimed at the protection of farmland. As the approaches to farmland preservation in Canada vary by province, this chapter provides the context for land use planning for agriculture in Canada and focuses on the two most populated provinces, Ontario and Quebec. It provides an overview of the Canadian context and a detailed assessment of what is occurring in Ontario, which uses a policy led system and Quebec, which uses a legislative system.

2.2 Farmland Loss and Urban Development: The Canadian Context

Farmland loss to urban development has been a relevant issue recognized in Canada since the late 1960s and especially the 1970s. This problem is mainly due to urbanization that, between 1971 and 2001, consumed about 15,200 km^2 of agricultural land (Connell et al. 2013). However, reconciling competing interests for agricultural lands remains a complicated process that crosses multiple jurisdictions and involves, under Canada's Constitution Act, federal and mainly provincial governments, as well as regional and local municipalities. In Canada, responsibility for agriculture is jointly shared by the federal and provincial governments (Canada's Constitution Act)—farmland conservation policies can be formulated by different forms of legislation or guidelines at different political levels but the principal levels are provincial and regional or municipal (Caldwell et al. 2017). The recent development of these policies, the most significant of which were put in place during the 1970s (Bryant and Russwurm 1979), reflected the recognition of several socioeconomic and environmental conflicts between various land uses, such as urban, recreational and industrial on the one hand, and rural or agricultural communities on the other hand (Hiley 2017).

Farmland preservation in Canada has become a major issue of public interest. However, the magnitude of the issue, as well as its recognition, varies across the country; in addition, under the Canadian constitution, each province has its own legislative authority to establish its own approach to preserve farmland and farming. Factors such as food production, market value for land, environmental issues, amenity of rural landscapes, agrarian ideals, land use conflicts in the urban fringe and loss of prime farmland has substantially influenced farmland policies in Canada, especially in Ontario (Wilton 2017; Hoffman 2001; Simpson-Lewis et al. 1979), as well as in British Columbia and Québec where farmland protection or conservation legislation was put in place during the 1970s at the provincial level (Bryant and Russwurm 1979, 1982).

It can be argued that the strongest policies regarding farmland preservation and urban development exist in Ontario, which is important, given current urban growth pressures. Since the 1970s, to cope with the issues of rapid urbanization, Ontario developed different tools aimed at preserving farmland, at various levels of government, with varying levels of success. Alternately, in 1978, Quebec sought to conserve good quality farmland through development planning, creating a legislative framework for the protection of farmland. In the mid-1990s, an important change was made to include the protection of farm activities, in effect recognizing that it is not possible simply to protect farmland from urbanization and other land use pressures through land use planning without paying greater attention to farming activities themselves. The approaches in Ontario and Quebec are quite different and provide strong case studies regarding farmland protection in Canada. In the following sections, an analysis of farmland preservation strategies in Ontario and Quebec is presented.

2.3 Ontario: Policy Responses to Population Growth and the Protection of High Quality Farmland

Ontario is the most populous province in Canada. The province has nearly 13.5 million inhabitants representing more than 38% of the country's total population. The largest metropolitan centre is focused on the City of Toronto and in 2016 the City of Toronto and the surrounding four regions had a population of nearly 6.5 million residents. In fact, the *Greater Toronto Area* (GTA) is one of the fastest growing regions in North America.

Ontario is also home to Canada's best farmland. While in 2016 the province included 7.8% of Canada's total area in agriculture, it generated 21.6% of the country's total farm cash receipts (Statistics Canada 2017). This reflects the combination of high quality soils (including more than 50% of Canada's Class 1 farmland) and the country's best climate for agriculture. This combination of soils and climate come together in the southern part of the province.

This mixture of significant population growth and high quality farmland is a recipe for pressures that have resulted in the on-going conversion of farmland to non-farm uses. Population forecasts suggest that these pressures will continue for generations to come, unless of course much more emphasis is placed on regional development in other regions of the province.

In response to the loss of farmland, the province of Ontario has pursued a policy-led system intended to slow down this pressure since the mid 1970s. Ontario's approach is in contrast to the legislative approach pursued in Quebec and British Columbia and is also in contrast to the authors' experience in the United States. The approach is best described as a "nested" policy-led planning system. Within Ontario, provincial legislation and related policy establish certain expectations for the planning system implemented through a coordinated approach. Provincial legislation and policy establishes a multi-tiered and interconnected framework to direct policy and decision making at the regional and local level. Within Ontario, land use planning decisions are primarily the responsibility of upper tier (including counties and regions) and lower tier municipalities (including cities, towns and townships). These decisions are, however, subject to provincial policy, which establishes the protection of farmland as a provincial interest. The result is that lower tier planning policy is nested within upper tier planning policy, which is nested within Provincial Policy. The result is that the provincial framework is reflected in municipal planning documents helping to ensure that provincial perspectives on planning for agriculture are reflected locally.

2.3.1 The Provincial Policy Instruments

2.3.1.1 The Planning Act and Provincial Policy Statement (PPS)

Within Ontario, the Planning Act provides the foundational framework for this nested policy-led system. The legislation directs municipalities and other government agencies to make planning decisions consistent with the provincial interest, including "the protection of the agricultural resources of the Province". These provincial interests are expressed in the Provincial Policy Statement (PPS). A revised *Provincial Policy Statement* (PPS) was released in 2014, although the province has encouraged policy to protect farmland since the mid-1970s (see for example the Foodland Guidelines that were in place from 1978–1994).

The PPS stablishes a number of policies connected to planning for agriculture. Some of the more salient features include:

- protection of specialty crop areas;
- direction that prime agricultural areas "shall be protected" for long-term use for agriculture;
- establishment of restrictive criteria for the expansion of settlement area boundaries in prime agricultural areas;
- the virtual elimination of the creation of new parcels for residential development in prime agricultural areas (limited only to "a residence surplus to a farming operation");

- allowance for the establishment of pits and quarries under certain criteria;
- support for agricultural uses, agriculture-related uses and on-farm diversified uses;

The PPS applies across the Province of Ontario and establishes a benchmark that is connected to other provincial initiatives. Two of the most important initiatives are reviewed in the following section.

2.3.1.2 Smart Growth in Ontario[1]

In 2004, the province of Ontario began to release discussion papers and proposed reforms to planning legislation developed by Smart Growth panels to assist the province, as well as regional and local governments, making decisions linking transportation, the natural environment, infrastructure, planning and public investment. The smart growth movement in Ontario was driven by the need for the provincial government to address several major planning problems (sprawl, gridlock, infrastructure, protecting the environment, and encouraging economic growth) that gained public profile in the 1990s and to provide specific plans for high growth regions of Ontario (Davidson 2017).

Davidson (2017) considers these suggested reforms and plans (smart growth) represent the most concerted effort to change planning directions in a generation and will guide planning in Ontario for some time in the future. Major components of smart growth in Ontario included:

- Planning Reform
- Large-Area Integrated Planning
- Infrastructure Renewal

The *Places to Grow Act* and the *Greenbelt Act* are both examples of the large-area integrated planning that were implemented to manage future growth in Ontario.

2.3.1.3 Greenbelt Act (2004)

The *Greenbelt Act* was implemented as law in 2004. The *Greenbelt Act* authorized the provincial government to designate a *Greenbelt Area* and develop a *Greenbelt Plan*. Both the Greenbelt Area and Plan were created in 2005 and updated in 2017.

[1]For further details connected to this section please see: Lot Creation In Ontario's Agricultural Landscapes: Trends, Impacts And Policy Implications authored by Wayne Caldwell, Claire Dodds-Weir, and Anneleis Eckert, School of Environmental Design and Rural Development, University of Guelph, 2012, available on: http://www.waynecaldwell.ca/Projects/ruralnonfarmrevisit.html (last access: 23/01/2018).

The purpose of the *Greenbelt Plan* was to permanently protect 1.8 million acres of farmland and natural environment surrounding what is referred to as the *Greater Golden Horseshoe* from urban development and sprawl. The *Greenbelt Plan* identifies «*where urbanization should not occur in order to provide permanent protection to the agricultural land base and the ecological and hydrological features, areas and functions occurring on this landscape*» (Ontario Ministry of Municipal Affairs and Housing 2017, p. 8). Figure 2.1 illustrates the extent of the lands impacted by the Greenbelt Plan in the *Greater Golden Horseshoe* in Ontario (including the GTA and neighbouring lands that wrap around the western end of Lake Ontario).

Lands within the *Greenbelt Plan* are referred to as "Protected Countryside". Within the Protected Countryside, land is identified as part of the Agricultural System, Natural System or Settlement Areas. Within the Agricultural System, land is either identified as specialty crop land, prime agricultural area or rural area, as identified in local official plans.

The Greenbelt Plan provides the following description of the Agricultural System: «*The Protected Countryside contains an Agricultural System that provides a continuous, productive and permanent agricultural land base and a complementary agri-food network that together enable the agri-food sector to thrive. Many of the farms within this system also contain important natural heritage features, including areas that support pollinators, and hydrologic features. The stewardship of these farms facilitates both environmental benefits and agricultural protection. The agricultural land base is therefore integral to the long-term sustainability of*

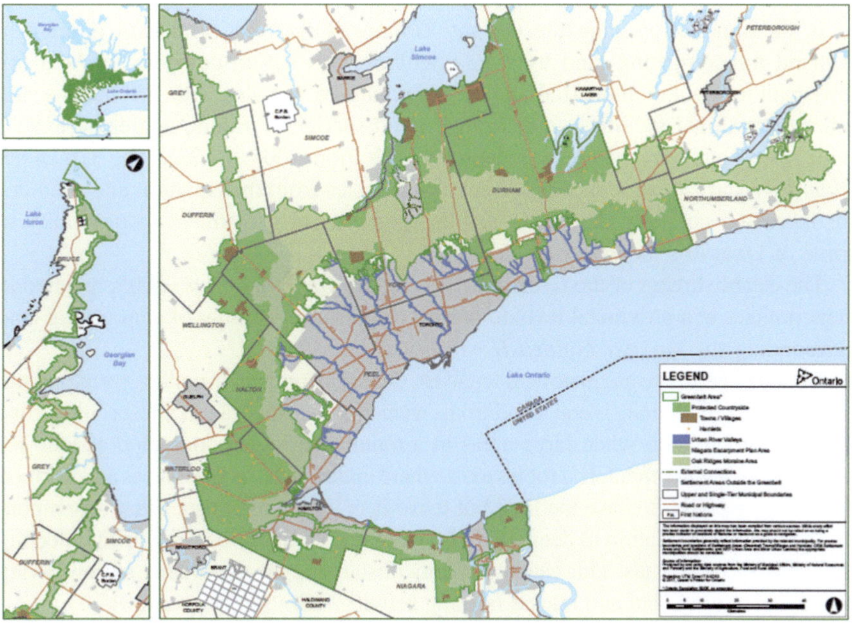

Fig. 2.1 Greenbelt Plan area. *Source* Ontario Ministry of Municipal Affairs and Housing (2017)

the Natural Heritage System within the Protected Countryside. It is through evolving agricultural and environmental approaches that this relationship can continue and improve» (Ontario Ministry of Municipal Affairs and Housing 2017, p. 15). The Greenbelt Plan, 2005, established policies on land use and lot creation within the Agricultural System.

The Provincial framework recognizes the importance of certain lands that have high capability for the production of tender fruit. These areas are identified as "specialty crop areas". These areas receive special protection. Urban areas cannot expand onto specialty crop areas, nor can specialty crop areas be re-designated for non-agricultural uses. Within prime agricultural areas, the following are permitted:

- normal farm practices and full range of agricultural, agriculture-related, and secondary uses;
- minor lot adjustment severances;
- farm split severances, provided severed and retained parcels are both 100 acres;
- surplus dwelling severances;
- severance for agricultural related uses.

Prime agricultural areas shall not be re-designated for non-agricultural uses except refinements to designations, or limited settlement expansion subject to the policies of the *Greenbelt Plan*. All new land uses shall comply with minimum distance separation formulae.

The vision of the *Greenbelt Plan* is to protect against the loss and fragmentation of agricultural lands, provide permanent protection to natural heritage features and water resource systems, and to provide for a range of economic and social activities associated with urbanization.

In the area surrounding the *Greater Golden Horseshoe*, rural land is often in demand for competing land uses. «Agriculture also has to compete for land not only with residential, industrial and commercial uses, but also with golf courses and other recreational facilities, transportation and service corridors, aggregate extraction, open space and wetland complexes» (Walton 2003, p. 23). The *Greenbelt Plan* has removed some of the competing land uses from the equation and provides for increased certainty that may facilitate agricultural investment in proximity to some of Ontario's largest urban centres.

The establishment of the *Greenbelt Act* and the *Greenbelt Plan*, 2005, resulted in clear policies at a provincial level to protect a significant amount of agricultural land surrounding the *Greater Golden Horseshoe*.

Nonetheless, there are historical legacies. On occasion land use plans were put into effect many years ago, when conditions were not as strict about preserving prime agricultural land and when large areas in a municipality were identified as land for subdivision development. Examples exist where under such circumstances a developer/speculator would purchase the land but leave it in agricultural use. Then, when conditions change in relation to demand for housing, the land may be developed reflecting these historic approvals. This can occur, notwithstanding the importance of conserving prime agricultural land and related farming activities (Bousbaine et al. 2017).

2.3.1.4 Places to Grow Act (2005)

The *Places to Grow Act* came into effect in Ontario in 2005. This legislation allows the government of Ontario to identify and designate growth plan areas and develop strategic growth plans for those communities.

Places to Grow is the Ontario government's program to plan for growth and development in a way that supports economic prosperity, protects the environment and helps communities achieve a high quality of life across the province. Growth plans assist the provincial government plan for growth and infrastructure investments across municipal boundaries.

The first growth plan was created by the Government of Ontario in 2006 for the *Greater Golden Horseshoe*. It guides decisions on how land is developed, resources are managed and public dollars are invested

The growth plan for the *Greater Golden Horseshoe*, 2006, established firm urban boundaries, requires residential intensification within those boundaries, and the creation of compact, complete communities. The *Greater Golden Horseshoe Growth Plan* identifies prime agricultural areas within the *Greenbelt Plan* and land use within those areas as being subject to those policies.

Places to Grow is primarily an initiative to manage urban growth and development. It is driven by the need to accommodate more population in an affordable way within existing urban areas. While it does not directly manage rural development, farmland preservation or non-farm development, it does create some benefits for rural areas by reducing the amount of agricultural land that is consumed by expanding urban uses and minimizing the effect of the spread of the urban shadow around Ontario's largest urban region.

The *Places to Grow Act* and its resulting plans are an urban expression of smart growth in Ontario. Davidson states «*while there are numerous benefits to rural areas from appropriate urban planning strategies, smart growth cannot be seen as a rural development strategy*» (Davidson 2017, p. 208). Davidson (2017) notes that if these plans are effective over the long term, smart growth can bring stability to the rural and agricultural land market that would benefit agricultural producers in Ontario.

As with the *Provincial Policy Statement*, the *Greenbelt Plan* and *Greater Golden Horseshoe Growth* Plan are largely implemented through the local and regional planning documents.

2.3.2 *Ontario's Planning System: Outcomes and the Protection of Farmland*

The *Planning Act* requires that the decisions of councils and other planning authorities "shall be consistent with" the PPS and "shall conform with" provincial plans or "shall not conflict with them" as appropriate. Provincial legislation establishes the authority for ensuring an appropriate level of consistency and resolving any conflict associated with interpretation or differing policies. For example, the *Greenbelt Act* prescribes that the policy direction afforded by the provincial *Greenbelt Plan* prevails in cases where there is conflict between the Plan and any local planning documents.

The overall result is that within Ontario, a relatively robust approach to the protection of farmland has been created. Municipalities are required to work with the provincial interest as expressed in several provincial planning documents. While this may be labeled as a top-down approach, it is required and widely applied, through local planning documents. It is also reinforced through provincial tribunals, provincial approvals and municipal planning decisions. While land continues to be lost, development is occurring at relatively high densities and is contiguous to existing urban settlements. The end result is a slowing of farmland conversion and the achievement of better planning outcomes.

2.4 Québec: Protecting Prime Agricultural Land and Integrating Strategic Development Planning for Agricultural Development

2.4.1 Québec's Legislative Basis for Preserving Farmland and Farm Activities

Québec was the second province after British Columbia to introduce legislation to protect prime agricultural land in 1978 and subsequently, reserves of prime agricultural land in the different *Municipalités Régionales de Comté* (Regional County Municipalities) (MRCs) were identified and created. Development pressures continued especially in the Montreal region. The *Commission de protection du territoire agricole du Québec* (CPATQ) subsequently played a key role in verifying requests to remove land from the agricultural reserves created in the MRCs (Choquette et al. 2013). However, some prime land was removed from some reserves, and the Provincial Government itself was able to substitute itself for the CPATQ and remove prime agricultural land from an agricultural reserve for the 'benefit of society'—e.g. to create an industrial park with the hope of supporting new employment development, or to create subdivisions. This meant that the protection legislation could be circumvented by the province itself.

Subsequently, in the middle of the 1990s, the farmland protection legislation was modified to include not just conserving the farmland but also protecting the farming activities that had developed on this farmland. This was the first step towards acknowledging the fact that land use plans, including protected agricultural reserves, could not in themselves conserve the prime agricultural land. However, there were still situations when prime agricultural land was removed from an agricultural reserve again 'for the benefit of society'.

2.4.2 Planning Agricultural Development in the Agricultural Reserves

This eventually led to important modifications in the way that agricultural land and its farming activities in the agricultural reserves were managed. This can be partly explained by the fact that the *Ministère de l'Agriculture et des Pêcheries du Québec* (MAPAQ) had begun to employ many young civil servants who had gone through an educational process in which concepts such as the multi-functionality of farmland and farm activities had been integrated. Thus, in 2008, the construction of *Plans de Développement de la Zone Agricole*—or *Agricultural Development Plans for the Agricultural Zone* (or Reserve) (PDZA)—were introduced, with each MRC being responsible for undertaking and managing this process and involving the appropriate range of other actors. The PDZA were in effect akin to a *Strategic Development Plan for Agricultural Development*, a development plan that necessarily had to take account of the different stressors faced by agricultures in the reserves, including initially the strengths and weaknesses of agriculture in the different reserves, competition from other regions and countries, as well as other stressors, including opportunities.

These PDZA converged with what is known as a *Sectoral Strategic Orientation* for agricultural development (Bryant 1995). In addition, the MAPAQ strongly suggested that in the process of constructing a PDZA the actors involved (farmers and their association the *Union des Producteurs Agricioles du Québec* (UPA) and their *Clubs Conseils*) should also take account of the multi-functionality of prime agricultural land and its farming activities and be prepared to involve other actors who are concerned with what might happen to these other functions. These other functions can include the protection of water quality, the conservation of heritage landscapes, and the conservation of rural tourism resources. One of the reasons for paying attention to these and other functions is that placing an emphasis on maintaining these other functions, actually supported by agriculture, can help mobilize the support of non-agricultural actor, including non-farm citizens. However, in reality some MRCs implicated other actors directly, while others did not.

Evidently, one of the most positive aspects of strategic development planning is that such a plan can deal with stressors (both opportunities and challenges) that cannot be managed through land use planning of any kind. For instance, there is growing concern in the markets of foodstuff for 'healthy' food produce and especially food produce that is produced locally. This opportunity can be dealt with in a strategic development plan through the initiation of different actions aimed at encouraging the production of healthy foodstuff. Another stressor that has been recognized in Québec particularly since the 1990s is climate change and variability and its effects upon agriculture. This phenomenon in some territories can create new opportunities for agriculture while in other territories it can also create major problems such as reducing the yields for some crops in some territories. In the case of the PDZA, since summer 2016, a major initiative was established to help provide farmers and their counsellors. This initiative has involved a committee that started work early in the fall of 2016 managed by the company *ÉcoRessources* and financed by *Ouranos* (the provincial organization that supports research projects and other initiatives dealing with climate change and variability).[2] The aim is to develop a set of tools that will help farmers adopt adaptation strategies to deal with the challenges created by climate change and variability. This effort is in its final stages since late fall 2017 and the next step will be helping farmers use these tools and adapt to climate change and variability, by focusing initially on one MRC.

2.5 Conclusion

Evidently, prime farmland represents a significant resource for the population for the production of appropriate foodstuffs. While Canada is a large country, Class 1 soils are relatively limited and mainly concentrated in the province of Ontario. Two provinces have enacted legislation to protect prime agricultural land (British Columbia in 1972 and Québec in 1978), while another province, Ontario, has focused on the utilization of various land use planning tools. The need to protect and conserve prime agricultural land (but also its farming activities) has been increasingly recognized in Canada, and at the same time, several jurisdictions have recognized the need to have broad support for protecting prime agricultural land and its farming activities, hence the increasing reference made to 'selling' the idea of the multi-functionality of prime agricultural land and its farming activities.

It has never been an easy process. In this chapter, we have dealt with at some length, the policy-led approach in Ontario and the trials and tribulations of this process in Québec. In Ontario, significant growth pressures have rapidly paved over significant amounts of the province's Class 1 soil. In recognition of the threat of future urban development on prime agricultural land, the provincial government created the *Greenbelt Plan*, protecting over 2 million acres of land. Challenges within the provincial planning system and delays between local land use planning approvals and development of the land have resulted in the continued loss of prime farmland, including areas that are now protected by the Greenbelt. While some development within the Greenbelt area has occurred, it is a result of historic planning decisions that predate the Greenbelt Plan. Given future growth pressures within the GTA and surrounding area of the *Greater Golden Horseshoe*, protection of prime agricultural land within southern Ontario is necessary. As local planning bodies are required to comply with provincial legislation, future land use decisions within southern Ontario are expected to limit the loss of prime agricultural land through smart growth initiatives.

Comparatively in Québec, it was recognized that land use planning tools alone cannot guarantee the protection of this valuable resource and the farming activities

[2]Bryant is a member of this committee.

it supports. Hence, it was eventually recognized that in order to really conserve prime agricultural land, one had to: a. support the development of sustainable agricultures and garner the support of the broader population; and b. use strategic development planning with its ultimate focus on actions and initiatives. Québec is one of the few jurisdictions in the world where this has all come together. This has been achieved ultimately by using strategic planning to encourage the development of sustainable agricultures and this in turn has depended upon a process in which citizens and various collective actors have been able to play a major role in the development planning process (although the extent to which this has occurred varies between territories or agricultural reserves), a phenomenon related to a major change in the way government has seen its role from top-down, to more of a role in supporting the developments proposed by other actors, including farmers, their associations and counseling clubs and citizens more generally. Despite all this, there is still much to be undertaken and to continue to emphasize development planning that involves the people who are most concerned by what happens to agriculture and its resources.

References

Agriculture and Agri-Food Canada. (2017). We grow a lot more than you may think. http://www.agr.gc.ca/eng/about-us/publications/we-grow-a-lot-more-than-you-may-think/?id=1251899760841#alt. Accessed November 1, 2017.

Bousbaine, A., Akkari, C., & Bryant, C. R., (2017). What can agricultural land use planning contribute to food production and food policy? *International Journal of Avian & Wildlife Biologyl, 2*(1), https://doi.org/10.15406/ijawb.2017.02.00009.

Bryant, C. R., & Russwurm, L. H. (1979). The impact of non-agricultural development on agriculture: A synthesis. *Plan Canada, 19*(2), 122–139.

Bryant, C. R., & Russwurm, L. H. (1982). North American farmland protection strategies in retrospect. *GeoJournal, 6*(6) (Special issue: Farmland preservation in North America), 501–511.

Bryant, C.R. (1995). Mobilizing and planning the community's strategic orientations: Basic tips. Sustainable Community Analysis Workbook 4, St. Eugène, Ontario: Econotrends Ltd.

Caldwell, W., Hilts, S., & Wilton, B. (Eds.). (2017). *Farmland preservation: Land for future generations*. Winnipeg: University of Manitoba Press.

Choquette, C., Guilhermont, E., Desjardins, M. C., Bryant, C. R., Gilles, D., Comtois, S. (Groupe de Recherche sur les Stratégies et les Acteurs de la Gouvernance Environnementale (SAGE)). (2013). *Étude de cas de la CPTAQ (Commission de protection du territoire agricole du Québec): Historique, fonctionnement, résultats.* Québec: Université de Sherbrooke.

Connell, D. J., Bryant, C. R., Caldwell, W. J., Churchyard, A., Cameron, G., Johnston, T., et al. (2013). Food sovereignty and agricultural land use planning: The need to integrate public priorities across jurisdictions. *Journal of Agriculture, Food Systems, and Community Development, 3*(4), 117–124. https://doi.org/10.5304/jafscd.2013.034.011.

Connell, D. J., Caldwell, W., Bryant, C., Cameron, G., Johnston, T., & Margulis, M. (2016). *Farmland: A prerequisite for farmers*. Food and Agri-food Policy: Policy Brief. https://doi.org/10.13140/RG.2.2.15983.43689.

Davidson, G. (2017). Smart growth in Ontario: Getting ahead of the future. In W. Caldwell, S. Hilts, & B. Wilton (Eds.), *Farmland preservation: Land for future generations* (pp. 114–123). Winnipeg: University of Manitoba Press.

Hiley, J. (2017). Opening gates, not reinforcing fences: Land use planning in Canada. In W. Caldwell, S. Hilts, & B. Wilton (Eds.), *Farmland preservation: Land for future generations* (pp. 94–113). Winnipeg: University of Manitoba Press.

Hoffman, N. (2001). Urban consumption of agricultural land. In *Rural and Small-Town Canada Analysis Bulletin*, 3(2) (September 2001) (Catalogue No. 21–006-XIE), Ottawa: Statistics Canada.

Ontario Ministry of Municipal Affairs and Housing. (2017). Greenbelt plan. http://www.mah.gov.on.ca/Page13783.aspx. Accessed November 1, 2017.

Ontario Ministry of Agriculture, Food and Rural Affairs. (2017). Number and Area of Census Farms, Canada and the Provinces, 1996, 2001, 2006, 2011 and 2016. http://www.omafra.gov.on.ca/english/stats/census/number.htm. Accessed November 1, 2017.

Simpson-Lewis, W., Moore, J. E., Pocock, N. J., Taylor, M. C., & Swan, H. (1979). Canada`s special resource lands: A national perspective of selected land uses. *Map Folio*, No. 4, Lands. Ottawa: Directorate, Environment Canada.

Statistics Canada (2017). *2016 census of agriculture*. https://www.statcan.gc.ca/eng/ca2016. Accessed November 1, 2017.

Walton M. (2003). Agriculture in the central zone. Neptis Foundation. http://www.neptis.org/sites/default/files/smart_growth_issue_papers_agriculture/agriculture_nip1.pdf. Accessed November 1, 2017.

Wilton, B. (2017). Farmland preservation perspectives. In W. Caldwell, S. Hilts, & B. Wilton (Eds.), *Farmland preservation: Land for future generations* (pp. 1–12). Winnipeg: University of Manitoba Press.

World Bank. (2014). Agricultural Land (% of Land Area). https://data.worldbank.org/indicator/AG.LND.AGRI.ZS?locations=CA. Accessed November 1, 2017.

Sara Epp is a Ph.D. Candidate in the Rural Studies program at the University of Guelph. Sara completed her undergraduate and master's degrees in Geography at Brock University. She is currently a sessional lecturer in the Landscape Architecture and Geography programs at the University of Guelph. Growing up on a family farm in an agriculturally-focused community impacted Sara's research focus and academic pursuits. Her current research interests include rural land use planning and their impacts on farm viability, agricultural diversification and farmer resilience. Her dissertation will explore the expansion of agriculture in northern Ontario, focusing on the impacts of policy on farmer resilience and community economic development.

Wayne Caldwell is a Professor in Rural Planning in the School of Environmental Design and Rural Development at the University of Guelph. He has previously served in interim positions as Dean of the Ontario Agricultural College and as Associate Vice-President Research. His interests include planning for agriculture, farmland preservation and community-based approaches to economic and environmental issues. Recent books include: Planning for Rural Resilience and Farmland Preservation: Land for Future Generations. His 11th book will focus on planning for agriculture. He has served as Chair or President of a number of local, provincial and national organizations.

Christopher Bryant has a Ph.D. from the London School of Economics and Political Science (1970). He was a Professor of Geography, University of Waterloo, Ontario, from 1970 to 1990, Chair of the Department (1980–1983), and director of the Program for Local Development, a Canada-wide program in partnership with the Canadian Association of Economic Development, from 1984 to 1990. He was Full Professor of Geography at the Université de Montréal from 1990 and Adjunct Professor since September 2014. He is also Adjunct Professor with the School of Environmental Design and Rural Development, University of Guelph, since September 2012. He is an internationally recognized leader in research on rural areas and small towns, as well as the planning and management of periurban agriculture in North America and Western Europe. He is one of Canada's leaders in the adaptation of human activities to climate change and variability (CCV), notably agriculture for the last 29 years, and community resilience to CCV. His most recent book explores agricultural adaptation to climate change.

Chapter 3
Managing Urban and Rural Agricultural Landscape Processes in Japan

Chika Takatori, Nobuko Kawaguchi and Hiroyuki Shimizu

3.1 Change of the Positioning of Agriculture in Urban and Rural Areas

3.1.1 Positioning of Farmland at the Time of the City Planning Act (1968) and the Act on Establishment of Agricultural Promotion Regions (1969)

In Japan, urban agriculture historically developed in relation to the life of urban inhabitants, and declined when competition for urban land use emerged. In the *Edo* era (1603–1868), there was a circulation of resources between urban and rural areas. For example, raw sewage from the urban area was utilized as manure by suburban farmers, and the farm products were consumed by urban inhabitants. After the modernization period, however, such mutually beneficial relationships were lost; artificial manure became widely used in agriculture, and the demand for the development of houses intensified, resulting in competition for urban agricultural land. During the period of rapid economic growth from 1954 to 1973, residential land demand increased dramatically, and public opinion on the housing shortage led to strong criticism of the large amounts of land used for urban agriculture. Through the expansion of urban land use, farming was exposed to a headwind; thus, urban agriculture declined due to friction among neighbouring inhabitants caused by pesticide dispersion or dust, as well as the increase in the landholding costs brought on by the rise in land prices.

During such population expansion, the goal of the land use policy, which was established in the *City Planning Act* in 1968 and the *Act on Establishment of Agricultural Promotion Regions* in 1969, was to divide the urban and rural areas

C. Takatori (✉) · N. Kawaguchi · H. Shimizu
Graduate School of Environmental Studies, Nagoya University, Nagoya, Japan
e-mail: takatori@corot.nuac.nagoya-u.ac.jp

E. Gottero (ed.), *Agrourbanism*, GeoJournal Library 124,
https://doi.org/10.1007/978-3-319-95576-6_3

through zoning in order to prevent the expansion of residential areas in a disorderly manner on the outskirts of a city. Urban agriculture was regarded as, so to speak, a transient existence in both city planning and agro-politics. In the *City Planning Act*, the area division system was founded and associated taxation systems were revised. In the area division system, *Urbanization Promotion Areas* were set as "those areas where urban areas have already formed and those areas where urbanization should be implemented preferentially and in a well-planned manner within approximately the next 10 years", and *Urbanization Control Areas* were set as "those areas where urbanization should be controlled". Farmland in the *Urbanization Promotion Areas* could be freely converted to residential land if the applications from landowners were accepted in the *Agricultural Land Act*, and if the appropriate taxation system measures were applied.

On the other hand, the *Act on Establishment of Agricultural Promotion Regions* was created in order to preserve and form the agricultural areas, which were mainly composed of superior agricultural land, by saving the farmland through the control of the chaotic urban sprawl. In this Act, the *Agricultural Use District Area* was set as 'those areas that should be used only as agricultural land', and the conversion of the farmland was not admitted as a general rule. The measures for agriculture promotion, such as land improvement projects, were promoted systematically and intensively mainly in these *Agricultural Use District Areas*, and few measures were promoted in the agricultural fields in the *Urbanization Promotion Areas* (Fig. 3.1).

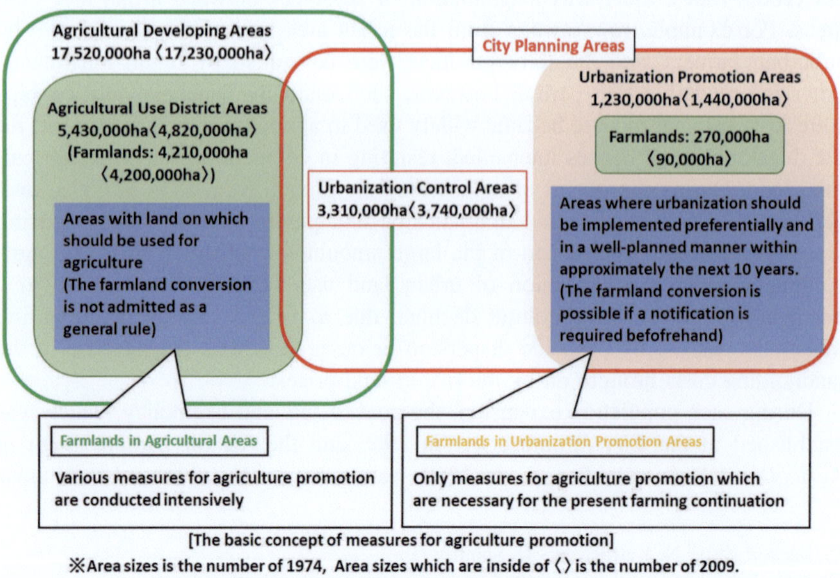

Fig. 3.1 The relationship between farmlands in *agricultural areas* and farmlands in *urbanization promotion areas* (in 1974). *Source* Ministry of Agriculture, Forestry and Fisheries (2011)

3.1.2 Revision of the Productive Green Space Act (1991)

During the late 1980s, land prices soared in the three major metropolitan areas in Japan and farmland in the *Urbanization Promotion Areas* was in strong demand to be converted to residential areas. To cope with this demand, the land was divided into "the farmland to be preserved" and "the farmland to be developed" by city planning in 1991. "The farmland to be developed" received the same municipal property tax rate as residential land, and the inheritance tax payment postponement system was not applied. On the other hand, "the farmland to be preserved", which is defined as "Productive Green Space", was imposed the same municipal property tax rate as agricultural land and the inheritance tax payment postponement system was applied in the revision of the *Productive Green Space Act* in 1991.

The requirements for designating farmland as *Productive Green Spaces* were (1) the farmland had to have a considerable effect on securing a good living environment and were suitable as a future public accommodation site, (2) Total area of the farmland was more than 500 square meters, and (3) The farmland was admitted with the understanding that the continuation of agriculture is possible depending on the condition of the draining system, and so on. The regulations pertaining to the *Productive Green Spaces* included (1) The land owner was obliged to manage the land as farmland; (2) Construction was permitted only for facilities which are necessary for agriculture, forestry, and fishery; (3) The farmland could be purchased by the municipality in the case that the main farmer was not able to engage in agriculture due to death, or after the farmland was designated as *Productive Green Spaces* for 30 years; and (4) When the transfer of proprietary rights was not carried out within three months from the day of the purchase proposal, act restrictions were cancelled. According to these provisions, the farmland categorized as *Productive Green Spaces* have been preserved; however, all other areas of farmland continued to decrease. The total amount of Farmland designated as *Productive Green Spaces*, and subsequently protected from urban sprawl, amounted to 16.3% of the total farmland (Figs. 3.2 and 3.3).

There are several problems with the system of *Productive Green Spaces*. First, the system was utilized in the three major metropolitan areas, but it is hard to say that the designation and preservation of the *Productive Green Spaces* represented an ideal urban vision for the future, as the district designations were dependent upon the landowner's intention. This is one of the reasons that municipalities seldom accepts the proposal of purchase by landowners. As this demonstrates, it was a good attempt by the Tokyo metropolitan government to enact a "district system to improve agriculture scenery". In this case, the government specified the area where farmland should be preserved in cooperation with local community planning; therefore, the purchase of the farmland improves the local environment, which is especially important for the promotion of the agricultural experience, the preservation of the natural landscape, and the development of public land into spaces such as urban parks.

Area sizes of Farmlands in the Urbanization Promotion Areas

	Three major metropolitan areas	Other cities	Total Amounts
Farmland aside from Productive Green Spaces	14,562ha (16.6%)	58,940ha (67.1%)	73,502ha (83.7%)
Productive Green Spaces	14,278ha (16.3%)	61ha (0.1%)	14,339ha (16.3%)
Total Amounts	28,841ha (32.8%)	59,001ha (67.2%)	87,841ha (100.0%)

Only short-term agriculture measures are taken (Farmland aside from Productive Green Spaces)

Agriculture measures (not only short-term) are able to be taken (Productive Green Spaces)

Fig. 3.2 Area sizes of farmlands in urbanization promotion areas in 2009. *Source* Ministry of Agriculture, Forestry and Fisheries (2011)

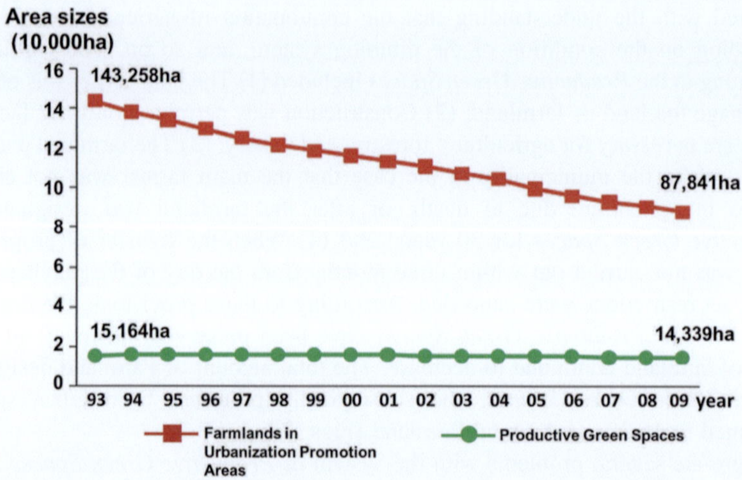

Fig. 3.3 Change of area sizes of farmlands in urbanization promotion areas. *Source* Ministry of Agriculture, Forestry and Fisheries (2011)

The second issue is that the designation of *Productive Green Spaces* is limited to farmland that is more than 500 m^2 in total area; thus, small farms under 500 m^2 are not eligible for preservation even if the landowners are willing to continue farming. This restriction also states that if one part of the *Productive Green Space* district is cancelled due to public expropriation or inheritance, then the entire district is subsequently cancelled if the remaining land area falls under 500 m^2, even if the landowner wishes to continue faming. Thus, appropriate measures should be taken for small farms under 500 m^2, as well as the *Productive Green Spaces* which turn into less than 500 m^2 against the intention of farmland owners, in order to maintain their relationship with the promotion of agriculture. Moreover, one of the restrictions of the current system of *Productive Green Space* is that a municipality can create a proposal of purchase once 30 years have passed from the original designation as a *Productive Green Space*. Therefore, the collateral quality of the urban farmland is weak.

The third issue is that the area designated as *Productive Green Spaces* in cities outside of the three major metropolitan totals less than 100 ha. This is because the forced continuation of farming for 30 years is a burden to the farmers, and the municipalities do not want to promote the designation, as this decreases the taxes, when the urban farmland is not positioned to be a major measure for agriculture promotion. Additionally, the preservation of farmland in *Urbanization Promotion Areas* is dependent upon the landowner's willingness. However, particularly outside of the three major metropolitan areas where the future residential demand is small, it is a concern that failed farmland may become unused land, a parking lot, or a material depot, and adversely affect the neighbouring living environment. In such cities, not only has the realization of the compact city been recently promoted due to population decline through the intensification of urban functions and residential areas, but also the measures of farmland preservation and urban agriculture promotion prove to be important for the formation of a positive living environment (Panel on Infrastructure Development 2009). It is important to consider the land use planning system and green master plan for the appropriate preservation of urban agriculture in order to meet the needs of urban farmers and the ever-changing circumstances of urban agriculture.

3.2 Current Situation of Urban and Rural Agriculture

3.2.1 Problems in the Periods of Population Decline

As previously mentioned, the purpose of such urban planning and agricultural policy was to divide the urban and rural areas through zoning in order to prevent the expansion of residential areas in a disorderly manner. In *Urbanization Promotion Areas*, the farmland was permitted to be developed. Only the "Productive Green Spaces" were designated as "the farmland to be preserved", and the measures for agriculture promotion were not promoted since the sites were assumed to be a future public accommodation site. On the other hand, in *Urbanization Control Areas* and *Agricultural Developing Areas*, farmland was assumed to be preserved and the various measures for agricultural promotion such as land improvement projects were conducted.

However, many researchers have pointed out that various problems have occurred today. First, the urban and rural divisions have failed; urban sprawl has expanded into urban periphery areas both in *Urbanization Promotion Areas* and *Urbanization Control Areas* due to loose land-use regulations and the excessive designation of the *Urbanization Promotion Areas*. Agricultural fields became small-scale and scattered due to the expansion of the city area and the occurrence of inheritance. Urban farmers have continued farming despite the various problems associated with physical environmental aggravation, such as the lower quality of water, sunshine and ventilation conditions, and the friction with neighbouring inhabitants caused by pesticide dispersion or dust. Furthermore, farmers of *Productive Green Spaces*, who are aging and have short of inheritors, cannot expect further farming in 2022 because it has been 30 years from the land's designation as a *Productive Green Space* under the revision of the *Product Green Spaces Act* of 1992. After the *Productive Green Spaces* are cancelled, the farmland will be converted into residential land and the sprawl will progress. On the other hand, when agricultural land is no longer needed to be converted into residential land due to a shrinking population, the remaining farmland will be scattered in a disorganized manner in urban areas and the abandonment of farming will progress. Similarly, in *Urbanization Control Areas* new development is limited and the aging farmers are experiencing a lack of successors, resulting in severe conditions.

Second, economic and environmental problems have been focused solely on farmland in the *Urbanization Control Areas* and *Agricultural Developing Areas*. As economic problems emerge, the agricultural prices fall and the packing cost and material price of fuel rise in accordance with the competition between production areas and the expansion of import volume; thus, the profit of farmers has been decreasing. The Japanese calorie-based food self-sufficiency ratio, which used to be approximately 80% in 1960, has continued to decrease to 39% in 2017. The impact of environmental problems such as the loss of ecological diversity has become an issue caused by excessive land improvement projects. New systems for solving these problems are needed today.

3.2.2 Expectation of Urban Agriculture by Urban Inhabitants and Enhancement of Citizen's Awareness of the Environment

On the other hand, urban residents have recently shown more interest in food safety and in urban and suburban agriculture. Fresh vegetables are more appreciated, and the availability of reliable farmland provides security for residents; therefore, the need to raise ones own farm products has increased. Furthermore, urban farmlands are regarded as valuable green spaces supplying a good living environment and refuge during times of disaster due to the expanse of a slow lifestyle, an increased sense of values, and improvement of the disaster prevention awareness caused by the Great East Japan Earthquake in 2011 (Fig. 3.4). The expectations of the urban agricultural fields has increased and include not only provisioning services, but also

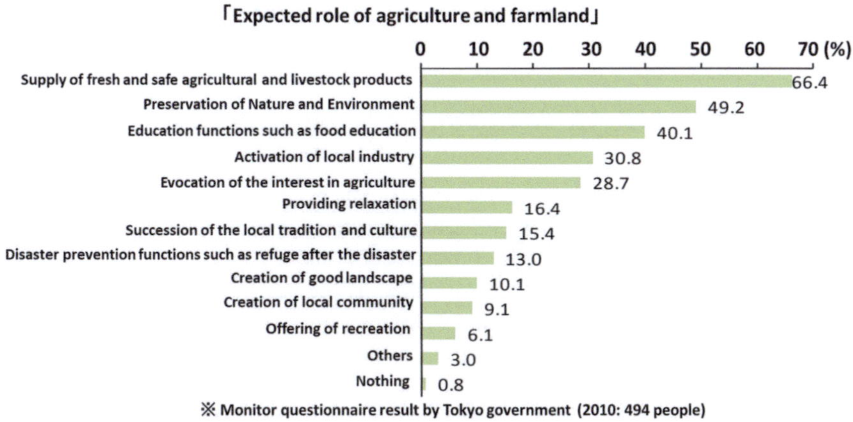

Fig. 3.4 Expectation of agriculture and farmland by urban residents by Monitor questionnaire by Tokyo government in 2010. *Source* Ministry of Agriculture, Forestry and Fisheries (2011)

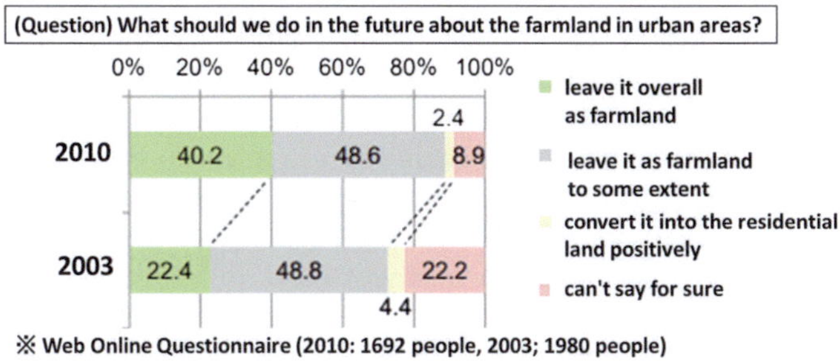

Fig. 3.5 Future expectation of urban farmland by urban residents by Web online questionnaire by Adachi-Ku in Tokyo government in 2003 and 2010. *Source* Ministry of Agriculture, Forestry and Fisheries (2011)

regulating and cultural services such as disaster prevention, environmental improvement, agricultural experiences, and landscape preservation. Additionally, new relationships between urban residents and urban farmers have been generated through a wide range of outreach attempts, such as participation in school education by volunteers and the management of an agriculture experience farm by farmers. Research has shown that urban residents favour leaving the space as farmland at least to some extent over conversion to residential land (Fig. 3.5), which further substantiates the positive relationship.

3.3 Recent Measures for the Promotion of Urban Agriculture

3.3.1 Established Act and Plan for Urban Agriculture

The *Urban Agriculture Promotion Basic Plan* was established in 2015 (Ministry of Agriculture, Forestry and Fisheries 2016a), and the *Urban Green Space Law* was revised in 2017 (Ministry of Land, Infrastructure and Transport 2017) in order to address the current urban agriculture situation. These new act and plan require that the duties of national and local public entities, agricultural organizations, urban farmers, and urban residents should be clear, and all stakeholders should cooperate with each other and take measures for the promotion of urban agriculture. The characteristics of urban agriculture demonstrate the potential for profitable farming due to diversified small-quantity production, direct marketing to urban inhabitants, and direct trade to companies; additionally, urban residents can deepen their understanding of agriculture, which is an invaluable tool. When we think of urban farming in this way, we realize that it is necessary for urban agriculture to change directions so as to be supported by the main agriculture promotion measures in agro-politics. Additionally, the direction of urban policy has shifted from meeting demand for urban land use during city expansion caused by population growth, to aiming for a 'compact city' during times of shrinking population and in the face of financial limitations from public bodies, in order to maintain the convenience and vitality of the cities. Urban farming is an important factor in the dynamics of the compact city. The new act and plan are the beginning of the policy shift, and address the following main points (Ministry of Land, Infrastructure and Transport 2017).

1. *To evaluate the mixture of urban and rural areas positively by regarding urban agricultural fields not as "the place to be developed" but as "the place to be preserved"*. In order for urban farmland to coexist with other urban land uses, and to utilize the various functions of urban agriculture, the positioning of urban farmlands in land use planning should be transformed. Additionally, it is necessary to preserve urban farmland premeditatedly. In the revision of the *Urban Green Space Law*, the following four points were changed or created.

    1.1 *Specification of "farmland" to the definition of "Open space"*. In the definition of "Open space" in the *Urban Green Space Law*, "farmland" has not been clearly specified. For example, only farmland which occurred in the forest was included in "Open space". However, in the revision the of Urban Green Space Law, the 'farmland', which is important for the creation of a good living environment, was specified in the definition of 'Open space' and it became the target of the various systems in the *Urban Green Space Law* (Table 3.1). For example, various green space preservation systems are able to be applied to the farmland that should be preserved for the prevention of disordered expansion of built-up areas as well as disasters, that

Table 3.1 Farmland use which corresponds to the definition of "Open Space" in Urban Green Space Law

Green form that is defined as "Open Space"		Corresponding farmland use
Woodland		Bamboo forest, Prum-grove
		Tea plantation
		Orchard
Grassland		Vegetation field
		Lawn
		Grazing land
Riparian place	Ponds and swamps	Lotus root, water shield
Rock place	–	–
Similar place	Wet zone	Paddy field
		Horseradish
	Sand dune	Pickled shallot, Melon

Source Ministry of Land, Infrastructure and Transport

become the habitat of rare creatures such as lightning bugs, and that contribute to the creation of improved landscapes such as terraced paddy fields or tea plantations. Corresponding to the revision of the definition of open spaces, farmland should be positioned in the *Green Master Plan*, which is created by each municipality (the creation of the *Green Master Plan* by municipality is a not duty). It is desired that this *Green Master Plan* be created through an agreeable formation process, and clearly show the role of urban agriculture and urban farmland in relation to the local residents.

1.2 *Restriction Alleviation of Productive Green Spaces.* The restrictions on Productive Green Spaces were eased in the revision of the *Urban Green Space Law*. First, the area of *Productive Green Space* for a group of some land owners could be brought down from over 500 m^2 to over 300 square meters if the municipalities establish ordinances. Secondly, the period when the purchase proposal of the *Productive Green Space* is enabled was postponed every 10 years if the designation of the *Productive Green Spaces* is over 30 years old. Thirdly, the establishment of direct sale locations or farmhouse restaurants was enabled within the *Productive Green Spaces.* These restriction alleviations were set in order to support the urban farmers who are willing to continue farming in the urban area.

1.3 *Rural residential districts are newly created in a zoning system.* In Japanese city planning, a zoning system that regulates the type of buildings to be erected in an area, called *Land-Use Zones*, was established in *Urbanization Promotion Areas*. A new type of district called "Rural residential districts" was created with the purpose to protect the positive living environment with low rise dwellings which harmonize with agriculture. In these rural residential districts, the area ratio of buildings, maximum height of buildings, and the location of walls are restricted in city planning. Additionally, the site of the building, structure, the building facilities, and the building use is restricted in the *Building Standards Act*. The significance of the creation of the rural residential districts can be summarized by the following two points. First is the introduction of building regulations in the farmland. In the *Urbanization Promotion Areas*, there were no regulations stating that the farmland, aside from the *Productive Green Spaces*, could not be developed as residential land. Under the new regulations established for buildings in the rural residential districts, people must receive permission from mayors, town managers, and village headmen when the form of the farmland is changed and the construction of a building or other establishment is conceived. Mayors should only permit the cases in which the size of the development is smaller than the specifications found in the ordinance so that it does not disturb the promotion of the convenience of agriculture and the protection of the positive living environment. Second is the alleviation of zoning restrictions that enable the construction of facilities for agriculture. In the current zoning system, the creation of agricultural facilities is not permitted in the exclusive residential districts, which encompasses relatively large amounts of farmland, according to the restrictions of the *Building Standards Act*. Permission from the specific administrative agency is necessary in order to build such facilities. On the other hand, zoning regulation permits the construction of agriculture facilities, farm restaurants, and farm product direct sales stores locations in the rural residential district.

2. *To promote further entry of various motivated entities into agriculture such as urban residents, and companies which promote new agricultural businesses.* In urban agriculture, it is easy to find a successor because of the prime location of the farmland; farmers benefit from stable income and get to enjoy the city life, culture, and entertainment. Urban farming should be family-owned so as to include successors; however, the aging population and declining birth rate will cause such urban family-owned farms to face difficulties. Therefore, in case the landowners are unable to continue farming themselves, it is important to find new farm operators through a loan of the land so that there is a stable continuation of urban agriculture. The new operators of the farm can be determined from the following options.

   The first alternative is farmers within the local districts who have previously farmed or people who have a willingness to farm. In this case, public bodies such as the agriculture committees of local municipalities are expected to take an active part in the intermediation. Local bodies and agricultural cooperatives can provide training in farming techniques and guidance in securing income for young people who desire to become new farmers.

   The second option is for food-related business operators to produce and sell the food through the establishment of an eating out service in the local area. These operators can collaborate with urban farmers and participate in farming by leasing the farmland from the owners. Contributions to the regional economy are expected by utilizing their knowledge, techniques, and acquaintances, and developing new products using local, fresh, and traditional vegetables. The public bodies, including the agriculture committee, should relay the needs of the diversifying citizens to the private enterprises which are available for them depending on their ability.

The third option is venture companies working in welfare, education, and IT. Until now, the local municipalities and agricultural cooperatives have taken a central role for the creation of opportunities for the citizens to experience farming by opening the community farms, resulting in the vitalization of the local community and improving the citizens' lives. However, the needs of city inhabitants have recently diversified, and some companies are growing rapidly by capitalising on the farming experience. These companies become popular because of their careful farm management and instruction, and enhance their presence as an alternative to the publicly managed community gardens.

The most important aspect is that the operators become industrious farmers and urban agriculture maintains a stable presence utilizing its various functions. Furthermore, some recent attempts have been made from urban residents such as the elderly who are interested in learning agricultural techniques; many have decided to work as farm volunteers, and groups working on local community revitalisation efforts utilize farmland when the manager is absent for a certain period of time. Matching these types of people with the farmland should be considered a practical solution to managing farmland under these circumstances.

3.3.2 Future Challenges and Perspectives of Urban Agriculture and Urban Farmland

The future challenges of urban farmland in the *Urbanization Promotion Areas* and *Urbanization Control Areas* lies in collaborating with the housing policy. With the developing of good housing stocks, the farmland should be divided into that of open spaces and that of agricultural utilization, as both of them are important for the enhancement of the living environment (Terada 2016).

The farmland for open spaces should be close to the residential land so that the residents can enjoy farming. Moreover, the land set aside for urban agriculture is also important and the vacant land should be inflected as farmland. To make this a reality, establishing a district plan during city planning can ensure a good mixture of residential areas and farmland by setting appropriate regulations. The common use of farmland by all stakeholders except for landowners can be achieved by separating possession and use of the farmland and compensating a landowner through a lesser property tax (Ibid.). The community farm system, which is specified in the *Act on Promotion of Development of Community Farms* in Japan and defined as the group of small farmlands that urban inhabitants (usually more than 10 people) can borrow in a self-support purpose and use, can also be useful for this purpose.

The farmland for agricultural use should be a unity of agricultural fields. In this kind of farm, the urban agricultural business which capitalises on the proximity to the city should be managed. These farmlands have a role as vacant land for disaster prevention during times of need; some areas can open as experience farmlands with stores and restaurants. For these preservation farmlands, a conversion permission system of the *Agricultural Land Law* or a system to follow the *Agricultural Land Areas* should be considered. Moreover, it should be considered that the city farm co-op, which has plenty of assets from the profit of urbanization, organizes the urban green farmland preservation fund by getting the financial assistance of municipalities and civic participations, performing land trust, and through the purchase, lease mediation, and financing of the business in order to preserve the farmland as public (Takeuchi and Matsuki 1987). It is important to create a good environment in the land use mixture areas by forming the unity of farmlands. Through periodic reviews of the area divisions, it is desired that the united farmlands located where future population decline is predicted in *Urbanization Promotion Areas* are admitted into *Urbanization Control Areas* (this is called "Reverse drawing").

3.4 Recent Measures for Promotion of Agriculture in Rural Areas

3.4.1 Established Act and Plan of Agriculture

The *Basic Plan for Food, Agriculture and Rural Areas* was revised in 2017 by the Ministry of Agriculture, Forestry and Fisheries (Ministry of Agriculture, Forestry and Fisheries 2017a). This Plan provides guidance to advance the measures taken by the Japanese nation so that Japanese agriculture and agricultural villages can meet the social/economic changes over the next 10 years. The agricultural villages in Japan face the severe reality of aging farmers and the dilapidation of the farmland. On the other hands, there are new movements that primary producers' diversification into processing and distribution is promoted. The "Sixth Industry" means that primary industries producers works not only production of agricultural, livestock and marine products but also food processing industry (secondary industry) and distribution & marketing (tertiary industry) (Ministry of Agriculture, Forestry and Fisheries 2017b). The agriculture and forestry marine products and the food is exported to the foreign countries, and the "Rural district recurrence" is led by the youth. Under such a circumstances, some of the major measures for the promotion of agriculture in this revise are the following points.

1. *'Industrial policies' for the promotion of the growth of agriculture industries through the acceleration of structural reforms and the uptake of new demand for agriculture.*

    1.1 *Promotion of the sixth industry by the creation for new value.* In rural areas and agriculture villages, various resources exist such as agriculture products, biomass, natural energy, traditional culture, and so on. On the other hand, the farming enterprise faces critical conditions such as the price slump of agriculture products and the rise costs of materials. The sustainable development of agriculture needs not only the produce itself, but also the promotion of the sixth industry which utilizes various local resources, and the high added value of the agriculture by agriculture-commerce-industry cooperation. This also helps to promote the export of domestic farm products, to secure employment within the rural areas and villages, and improves income. Moreover, it is important to strengthen the bond between producers and consumers via direct sales from the production locations. According to the census data of 2015 by the *Ministry of Agriculture, Forestry and Fisheries*, the annual total sales from farm products through direct sale locations increased by 997,400 million yen (6.6%) from the previous year and occupied 50.7% of agricultural revenue. The processing of farm products rose 4.0% to 892,300 million yen and occupied 45.3%. A sightseeing farm rose 3.8% to 37,800 million yen and occupied 1.9%.

    As a case for the sixth industry, Kyoto Co., Ltd. in Kyoto City, Kyoto Prefecture (founded in 2002) produces the traditional vegetable "Kujo leek" on his own, processes this to a cut leek or leek oil in his factory, and sells them to ramen restaurants, supermarkets, and department stores. Through sixth sector industrialization, the sales amount spread to 590 million yen in 2012. The representative director, who has business experience in the apparel industry, grasped the needs of the ramen industry by using his know-how and created the new demand of *Kujo-leek* in the Kanto district where the dietary habits of the blue leek were few. Furthermore, the new value of 'freshness' was created by the low temperature management in the manufacturing process, washing labour such as the sterilization with the ozone water, and direct shipment of the cool delivery service at the distribution stage. The production and management of the *Kujo-leek* was controlled and the quality was ensured through the construction of a production centre shipment relay system, as well as the combination of three agriculture production centres in the Kyoto prefecture.

    To promote the sixth industry, the improvement of a system for supporting farmers appropriately by promoting various efforts for human resource development is important. However, many farmers find it difficult to locate the supply of a fund with high flexibility, secure a market route, develop an effective management method, and match with processing and sales companies. Agriculture, forestry and fisheries Fund Act for Innovation,

Value-chain and Expansion Japan (Fund Law) was established in 2012 in order to help these farmers and prepare the necessary supporting system.

1.2 *Reform of Farm co-op and agriculture committee for the first time in 60 years.* In the revision of the *Basic Plan for Food, Agriculture and Rural Areas* in 2017, the role for farm co-ops and agriculture committees, which have large impacts on farming in Japan, was reformed for the first time in 60 years (Ministry of Agriculture, Forestry and Fisheries 2016b). The basic concept of this reform is that the farm co-op is the cooperative organization which farmers establish voluntarily; the leading role in the farm co-op organization is farmers and the second is the local farm co-op. The problems with the current system of farm co-ops and the points of reform at this time are examined in the following points.

(a) *The relationship between farmers and farm co-op.* Currently, is it not recognized that the biggest mission of the farm co-op is the advantageous sale of farm products and procurement of production material; the 'non-profit rule' of the farm co-op causes a misunderstanding that it should not gain a profit. Farm co-ops had a perception that it is natural for the members to use a farm co-op. In some cases, the farm co-ops were disclosed by an unfair business method in violation of the Antitrust Law. The people who are used to the conventional business methods including the consignment sales method constitute directors of farm co-ops, and it is difficult to perform economic activities for the positive improvement of the agriculture income. In this revision, the farm co-op should consider a maximum increase in the agriculture income, and the farm co-op should realize high profitability through precise operation. This should be allotted for the reduction to a member of an association (business use quantity allotment) and investment in the future. The farm co-op must not force the business aspect on farmers, and the majority of the directors should be the authorized agriculture professionals as a general rule. Moreover, the youth and women should be positively assigned as the directors.

(b) *Sale of farm products and Procurement of the production materials.* Currently, the farm co-op takes almost no risk associated with the consignment sales and they mainly do the collection of cargo and shipment, and the products are often sold cheap. Moreover, the procurement of the production materials is often from the *National Federation of Agriculture Co-operative Associations* or the *Association of Economic Farm Co-op Society* because the farm co-ops belong to Japan Agricultural Cooperatives and they are often expensive. As a result, farmers have recently stayed away from farm co-ops. Today the system of the direct marketing places, that producers can sell their products directly to consumers and determine the price and amount of products, is expanding and it causes the collapse of the farm co-ops (Kikuzima 2017). In this revision, the farm co-op should do the consignment sale and extend the reclaim market route in order to increase return by taking the appropriate risks. The procurement of the production materials should be obtained from the most beneficial places.

(c) *Management of the funding and cooperative.* Currently, the farm co-op receives the same regulations as banks and credit associations because it runs a trust business under the name of the farm co-op itself. Since a farm co-op is a cooperative, people who are non-members have regulation. In this revision, the farm co-op should transfer the trust business to the Central Bank for Agriculture and Forestry and offer only a financial service as the agency. The farm co-op should separate the parts of local services and change the organization to a company or the consumers' cooperative.

2. *'Regional policies' for promotion of the preservation and fulfilment of the multiple functions of agriculture and the farm village.*

2.1. *Maintenance of the village function by 'the collection and the networking'.* Regional policies for the aging and shrinking population are written in this revision. In order to continue the function of local communities, the promotion of 'a small base' policy that networks the peripheral villages and gathers life service functions for the basic village. On the other hand, the correspondence to damage to the birds and beasts, the security of employment and improvement of income in local villages by the creation of new value of agriculture-related productions and the new biomass industry, and the production and use of renewable energy is also written. Furthermore, the sustainable development of agriculture fields for the fulfilment of multiple functions through an exchange between urban and rural areas was also mentioned.

2.2. *Promotion of direct payment systems of multiple functions.* For regional policies, the direct payment systems of multiple functions are established. In Japan, there are three kinds of direct payment systems. First is the direct payment system of multiple functions. The purpose of it is to preserve and fulfil the multiple functions of farmlands and to support restructuring through initiatives such as the farmland. The method is to financially support something that the local communities can work on in the basic maintenance of the resources such as farmland, agriculture water, and the farm roads and the activities to improve quality; for example, farmland maintenance by local communities such as removing the mud from the waterway and mowing the farm roads. Such joint activities work on environmental conservation, such as the slight repair of facilities for agriculture or scene formation. Second is the direct payment system of intermediate and mountainous areas, for which the purpose is to revise the disadvantaged condition of the agricultural production in the intermediate and mountainous locations such as sloped areas. Third is the direct payment system for the environmental conservation type of agriculture, in which the purpose is to pay the additional cost for the farming that positively effects the prevention of global warming and preservation of biological diversity.

3.4.2 Future Challenges and Perspectives of Agriculture and Farmlands in Rural Areas

As previously mentioned, the recent revisions aim to improve the measures for agriculture promotion, specifically for the improvement of economic and regional conditions in rural areas. However, the measures are not sufficient for balancing the economy and the environment. In Japan, the land improvement projects were promoted from the revision of the Agricultural Land Law in the 1960s in order to promote the efficiency of farming. The productivity of Japanese agriculture has greatly improved by the land improvement projects. However, in terms of environmental problems, the impact such as loss of ecological diversity and multi functionality has become a problem caused by excessive land improvement projects. The existing research clarified the relationship between the loss of biodiversity and land improvement projects in Japan (Murase 2016). For environmental compensation, appropriate systems such as the direct payment systems are needed. Specifically, the direct payment system for environmental conservation type agriculture is the desirable policy for the cost allocation. The land improvement district occupies the important position for the subsidies in the direct payment system in Japan. The value conversion of land improvement projects and biological preservation is needed. To evaluating the multi functionality by clarifying the ecosystem services that agricultural fields have are the future challenges. On the other hand, recent enhancement of environmental awareness, environmentally-friendly agriculture is branded and has a high value to urban citizens. As a case, in 2007 Sado city in the Niigata prefecture established the new blend of rice in order to preserve the feeding ground of the Toki, a special national natural treasure. These kinds of new attempts for the promotion of agriculture are important for the future of the sustainable development of rural agriculture.

3.5 Conclusion

In Japan, the goal of land use policy was to divide the urban and rural area by zoning in the expanding periods. However, today in the shrinking periods, the need of diversion of agricultural land to residential land is declining. On the other hands, agriculture both in urban and rural areas faces economic and environmental problems. This chapter introduced the laws and some recent attempts of Japanese agriculture from the following four points; (1) To evaluate the mixture of urban and rural areas positively by regarding urban agricultural fields not as "the place to be developed" but as "the place to be preserved", (2) To promote further entry of various motivated entities into agriculture such as urban residents, and companies which promote new agricultural businesses, (3) To improve the self-sufficiency rate by the high technology collection, and (4) To balance the productivity and the biological diversity by branding of the environmentally-friendly agriculture. For the future sustainable development of urban and rural agriculture, it is needed for each area to make short- and long- term plan in line with the local conditions.

References

Kikuzima, R. (2017). Spatial competition among farmers' markets [農産物直売所の空間的競争]. *Journal of Rural Economics, 88*(4), 357–369.

Ministry of Agriculture, Forestry and Fisheries. (2017a). *The basic plan for food, agriculture and rural areas*. Accessed on May 31, 2017 http://www.maff.go.jp/j/keikaku/k_aratana/.

Ministry of Agriculture, Forestry and Fisheries. (2017b). *The situation of the sixth industrialization*. Accessed on September 1, 2017 http://www.maff.go.jp/j/shokusan/renkei/6jika/attach/pdf/2015_6jika_jyousei-58.pdf.

Ministry of Agriculture, Forestry and Fisheries. (2016a). *The urban agriculture promotion basic plan*. Accessed on May 1, 2016 https://www.mlit.go.jp/common/001131132.pdf.

Ministry of Agriculture, Forestry and Fisheries. (2016b). *Revise of the cooperative marketing act*. Accessed on January 1, 2017 http://www.maff.go.jp/j/keiei/sosiki/kyosoka/k_kenkyu/pdf/1_nokyohou_kaisei.pdf.

Ministry of Agriculture, Forestry and Fisheries. (2011). *The situation of urban agriculture*. Accessed on October 1, 2011 http://www.maff.go.jp/j/nousin/nougyou/kentoukai/dai1/pdf/tosi_kento1_siryo2.pdf.

Ministry of Land, Infrastructure and Transport. (2017). *Plan for drawing up and revising relevant ordinances according to the revise of the urban green spaces law*. Accessed on May 1, 2017 http://search.e-gov.go.jp/servlet/PcmFileDownload?seqNo=0000158872.

Murase, Y. (Eds.). (2016). Case studies in flatland farmland neighborhood landscape complex units: Asami District. In H. Shimizu, & C. Takatori (Eds.), *Labor forces and landscape management* (pp. 233–261). Singapore: Springer.

Panel on Infrastructure Development. (2009). *Report of basic problems and directions of urban policy*. Accessed on June 26, 2009 https://www.mlit.go.jp/common/000043480.pdf.

Takeuchi, T., & Matsuki, Y. (1987). Ecology and economy of urban agriculture. *City Planning Review, 145*, 35–40.

Terada, T. (2016). Urban periphery planning: Concept to link urban and rural communities in the 21st century. In H. Shimizu & C. Takatori (Eds.), *Labor forces and landscape management—Japanese case studies* (pp. 381–390). Singapore: Springer.

Chika Takatori is an assistant professor at the Graduate School of Environmental Studies in Nagoya University. She got her Ph.D. degree from the Department of Urban Engineering in the University of Tokyo. Her research focuses on the urban planning, landscape planning and design, urban environmental assessment, and land-use policies and practices. She won encouragement award and annual excellence paper award of the City Planning Institute of Japan.

Nobuko Kawaguchi Ph.D. is a researcher of the Institute of Materials and Systems for Sustainability (IMASS), Nagoya University. Her major research fields are Urban planning, Landscape planning, Land-use policies and Environmental assessment. Her particular interest is the sustainable landscape management on depopulating and hyper-aged society.

Hiroyuki Shimizu is an Emeritus Professor at Nagoya University. His research focuses on Architectural planning, Urban planning and Landscape planning. His recent publications: (1) Bernhard Müller, Hiroyuki Shimizu (Eds.), Towards the Implementation of the New Urban Agenda, Springer, 2018; (2) Hiroyuki Shimizu, Chika Takatori, Nobuko Kawaguchi (Eds.), Labor Forces and Landscape Management, Springer, 2017; (3) Hiroyuki Shimizu, Akito Murayama (Eds.), Basic and Clinical Environmental Approaches in Landscape Planning, Springer, 2014.

Chapter 4
Cities Linked Through Food Trans-Boundaries: The Case of Singapore as an Agri-Pelago

Jessica Ann Diehl, Ching Sian Sia and Amanda Jennifer Chandra

4.1 Singapore's Food Security

Known as the "City in a Garden", Singapore boasts lush public parks and buildings draped in vertical and rooftop gardens. The small city-state, with its multi-ethnic culture and influx of expats, overseas workers, and tourists, is also recognized internationally as a "foodie" destination, offering a dizzying array of authentic and fusion dishes from around the globe, and boasting 38 Michelin starred restaurants (including two hawker stalls; Singapore's equivalent to street food, but located in more hygienic open-air complexes). With a per capital income of $52,960.70 (US Dollar; 2016) (The World Bank 2017b), and a ranking of 4th place on the global food security index (GFSI; 2016) (The Economist Group 2017),[1] it would appear unlikely that Singapore would face a food crisis in the near future. However, a meager 0.9% of land use is allocated for agriculture, and farming is restricted to poultry, eggs, fish and vegetables (AVA 2017; The World Bank 2017b). As a result, Singapore relies heavily on overseas sources to sustain its food supply; an astounding 90% of domestic food demand is met through imports (AVA 2017).

Recent population estimates for Singapore are 5.7 million (Worldometers.info 2017) comprised of approximately 70% residents (citizens and permanent residents) plus 30% non-residents (other permit pass holders). An additional 16.4 million

[1]The GFSI ranks 113 countries based on availability, affordability, quality and safety of food. The index is a mix of quantitative and qualitative benchmarking models, constructed from 28 unique indicators, that measures these drivers of food security across both developing and developed countries. The GFSI looks beyond hunger to the underlying factors affecting food insecurity. Beginning in 2016 the GFSI includes an adjustment factor on natural resources and resilience. The new category assesses a country's exposure to the impacts of a changing climate; its susceptibility to natural resource risks; and how the country is adapting to these risks.

J. A. Diehl (✉) · C. S. Sia · A. J. Chandra
Department of Architecture, National University of Singapore, Singapore, Singapore
e-mail: akijac@nus.edu.sg

E. Gottero (ed.), *Agrourbanism*, GeoJournal Library 124,
https://doi.org/10.1007/978-3-319-95576-6_4

tourists were recorded in 2016 (Singapore Tourism Board 2016). With a land mass of 700 km^2, Singapore is categorized as 100% urban and boasts the third highest density in the world at 8155 p/km^2 (United Nations, Department of Economic and Social Affairs, Population Division 2016). Annually, Singaporeans consume 195,290 tonnes of chicken (35 kg per capita), 524,462 tonnes of vegetables (94 kg per capita) and 1,895,029,546 hen eggs (338 per capita) (AVA 2017). The paradox of Singapore's food security, ranking high on the global food security index, but relying on foreign food imports to meet the majority of food demand, is not unique. Rather, it is illustrative of a global trend in high urban, high-earning, net food-importing cities such as Saudi Arabia, Japan, the United Kingdom, Germany and the United States (Teng and Escaler 2010).[2,3] Singapore is more vulnerable to food availability than food affordability.

Can Singapore improve its self-sufficiency and reduce vulnerability in times of crisis? Globalization, climate change and urbanization increasingly have significant and deep impacts on local and global food systems; putting the country at risk for food shortages and price fluctuations beyond its control. Increasing competition for rural sources means that it is critical to consider potential changes in food demand by the urban populations in the top countries that Singapore imports food from. This chapter describes some of the social, political, and ecological factors that situate Singapore within an *agri*-pelago of cities linked through food trans-boundaries. First, we describe Singapore's move toward intensification and high-tech alternatives to producing food without rural land. Next, we consider Singapore's rural hinterlands that are geographically proximate but politically distant. We then turn to the major cities in top countries that Singapore imports food from to better understand competing urban demands including: Kuala Lumpur, Malaysia; Jakarta, Indonesia; and Sydney, Australia. Last, we look to China, where Singapore is leasing land for food production. We end with a discussion of Singapore's situation within an *agri*-pelago of cities linked through food trans-boundaries.

4.2 Land Scarcity and High-Tech Alternatives

Only 200 years ago, Singapore was covered in dense tropical forest surrounded by coastal mangroves. The island supported a tiny population of 1000 people dependent on fishing livelihoods (Swee-Hock 2012). With British arrival in 1819, the landscape changed rapidly as primary forests were cleared for large rubber plantations and settlements to accommodate the rapidly growing population, which

---

[2]Knoema—World Exports and Imports of Agricultural Products, https://knoema.com/cduhihd/world-exports-and-imports-of-agricultural-products, accessed 30 October 2017.

[3]United Nations Conference on Trade and Development, http://unctad.org/en accessed 30 October 2017.

reached 100,000 in 1871. At the time of independence from British colonization in 1965, 19% of the land area of Singapore was in agriculture. As an island with few natural resources and facing severe unemployment and a housing crisis, the newly established government of Singapore had a vision of modernization predicated on establishing a manufacturing industry, developing large public housing estates, and significant investment in public education. The ensuing decades witnessed rapid population growth and infrastructure upgrading. The current status as a world-class city is the result of that innovation, technological savvy, and financial investment.

As Singapore transitioned from a productive society to a consumer society, the landscape likewise transitioned from rural to urban. Between 1980 and 1989, farmland reduced from 12 to 3% of the total land area (Leitch Lepoer 1989; The World Bank 2017a). During that time, poultry and livestock farms were scaled down and there was a shift toward intensive farms that maximized production. This meant that crops like grains, pulses, root, fodder, and oil crops were impractical; Singapore became fully dependent on imported rice (Swee-Hock 2012). In 1984, the government decided it was more economical to import frozen and live pigs than to expend its budget on waste treatment of locally raised pigs—which produce five times more waste than humans; pig farms were phased out (Shim 1987). By 1991, less than 2% of Singapore's land area remained farmland. The domestic production of eggs had fallen to 35% of domestic demand and vegetables a measly 3%. In 2003, farmland was reduced to 1.2% with no domestic livestock. Currently less than 1% of the total land mass, farmland continues to shrink, albeit at a slower rate; and there is growing unease that purchasing power of food imports does not ensure national food security.

Singapore celebrated 50 years as an independent city-state in 2015, and the young nation has accomplished much. While arable land in Singapore has shrunk dramatically due to development, there is increasing interest in the sustainability of its food system and national food security. The government of Singapore has developed a food security road map with strategies that include (1) *food source diversification*; (2) *increase in local farm productivity of key food items*; (3) *stock-piling of food*; (4) *research and development* (R&D); and (5) *reduction of food waste* (Ludher 2015). R&D has instigated a fundamental shift in how Singapore endeavors to grow food: namely, through high-tech, high-producing, land-limited or land-less farms (see Box 4.1 for definitions). Public and private investment in R&D of high tech food production is taking the form of hydroponics, aquaponics, vertical and rooftop farming, and LED farming (see Box 4.2 for examples of R&D in Singapore). There are, however, critical voices, especially in the recent years, of insufficient evidence needed to confidently provide the recommendation to governments to promote urban agriculture (Badami and Ramankutty 2015; Warren et al. 2015).

Box 4.1 High-Tech Farming in Singapore

Hydroponics is the cultivating of plants without soil, using water with nutrients and minerals; typically, planted in a container with a gravel, sand or sawdust growing medium (Benton 2016; Resh 2002). It is a useful way of growing crops when arable land is limited. Crops are typically grown in a controlled environment, free of weeds and pests, and therefore healthier. By controlling nutrients, more plants can be grown per unit area, further resulting in greater yields. Plants also do not require change of medium or crop rotation like those grown in soil, where the fertility decreases over time.

Aquaponics combines aquaculture and hydroponics to raise fish and grow crops in a symbiotic environment. Aquaculture is the farming of fish, crustaceans, molluscs, aquatic plants, algae, and other aquatic organisms (FAO 2017). It involves cultivating freshwater and saltwater populations under controlled conditions. Aquaculture is a system were fish waste is used as an organic nutrient solution to grow crops; it is designed as a closed loop system that turns waste into resources. Using the soil-less hydroponic system of cultivating plants in a nutrient-rich water, it consumes less water and energy than traditional fish and plant crop farms (Goodman 2011). This system of fish farming and vegetable production requires very limited space. Like hydroponics, plants grown in an aquaponics system do not require chemical fertilizers because organic fish waste is sufficiently nutrient. Pesticides are also not needed in a controlled environment. The system is scalable and can be set up indoors or outdoors, allowing easy access to local fish and vegetable production.

LED farming is the growing of crops in an indoor environment with light-emitting diode (LED) lighting as the main source of radiant energy, simulating the photosynthetic waves of sunlight (Sabzalian et al. 2014). Red and blue light sources in LED lighting are able to maximize the absorption by chlorophyll, satisfying energy requirements of the plants (Mickens 2012). LED farming simulates photosynthesis by providing an adequate amount of photon fluxes. The lights can be switched on and off as required to maximize plant exposure but still simulating diurnal cycles, which produces significantly higher crop yields (Culzac 2014). LED lights consume less electricity and have a longer lifespan than typical fluorescent lighting. Research has also shown there were no apparent differences in plants grown under LED lighting and in a greenhouse (Sabzalian 2012). Plants grown in an LED farm also have a much lower water requirement, with claims of 1% of water demand compared to those grown outdoors.

Controlled Environment Agriculture (CEA) is farming in a controlled environment using hydroponics, aquaponics, LED farming. It ensures a stable supply of crops all year round, with plants protected from natural disasters and unpredictable weather such as earthquakes or droughts. As plants are grown in a highly controlled environment, there is also a lower risk of plants being exposed to plant diseases and insect pests.

Vertical farming produces food stacked in multiple layers, vertically inclined and/or integrated into a structure; it can be land-based or building integrated. Some vertical farms make use of CEA technologies, whereas others are exposed to environmental conditions. Due to the vertical orientation, vertically planted crops are exposed to less sunlight and often require supplemental lighting.

Rooftop farming occurs on the roof of a building or structure. In addition to producing food, rooftops planted with crops can provide other benefits including regulating building temperatures, reducing stormwater runoff, creating wildlife habitats. Rooftop farms can be planted using hydroponics, in containers, or in a soil-based or light-weight growing medium over a water-proof membrane.

Box 4.2 Research & Development

Rooftop farming. A study conducted by Ngee Ann Polytechnic students investigated the potential for an area in north Singapore to support rooftop farming. They calculated that 39,000 tonnes of hydroponically farmed vegetables could potentially be harvested from the 212 ha of vacant rooftops (Wilson 2005). A hydroponics demonstration farm was established on the rooftop of Singapore Changi General Hospital to provide healthy produce such as cherry tomatoes and herbs for patients (Wilson 2005). The hydroponics rooftop farm reduced the urban heat island effect of the bare concrete rooftop and eliminated the issue of sunlight glare into nearby hospital wards. Unfortunately, the hospital was unable to maintain the rooftop farm and it has been converted into a garden.

Hydroponic farming. In 2012, Singapore's first low-energy, hydraulic water-driven vertical farm, Sky Greens, was established. Sky Greens produces 2 tonnes of vegetables daily on its 3.65 ha farm using a vertical farming system known as A-Go Gro, jointly developed by Agri-Food and Veterinary Authority of Singapore and DJ Engineering Pte Ltd. (skygreens.com). The crop yield of certain vegetables is reported to be five to ten times the yield of traditionally cultivated vegetables (Wang 2012; Tan 2013).

Vertical farming. A 2012 study calculated that Singapore has 630–880 ha of public housing rooftop space with the potential for cultivation (Wong 2011; Tan 2013). If every public housing rooftop were to adopt the A-Go Gro system for crop production, the amount of food produced could meet 10–41% of total vegetable demand. Using the example of Kang Kong, a local variety of leafy vegetable particularly suited to the local climate and growing conditions, 60–200% of the demand for leafy vegetables could be met.

LED farming. In 2013, the first LED farm by Panasonic was established in northwest Singapore. The 248 m^2 indoor farm grows vegetables and root crops for domestic consumption including green lettuce, white radish, rocket lettuce, basil, mint herb, wild parsley, baby spinach, red lettuce, mini red radish, and potherb mustard (Manibo 2013). Crop yields are estimated at a growth of 2.5 times faster as compared to traditional farming methods, with a stable supply all year round. Vegetables are cultivated within 35 days in a vertical, soil-based method using LED lights, with the potential of vegetables and root crops to harvested within 28 days. The farm produces approximately 3.6 tonnes of vegetables each year and targets to meet 5% of the domestic vegetable demand in 2017.

4.3 Food Source Diversification & Competition

In addition to the figures on chicken, vegetables and eggs (which can be sourced locally), Singaporeans annually consume 300,000 tonnes of rice (53 kg per capita), 400,882 tonnes of fruit (71 kg per capita), 119,605 tonnes of fish and seafood (21 kg per capita), 114,220 tonnes of pork (20 kg per capita), and 32,966 tonnes of other meat (beef, duck, mutton) (6 kg per capita) (AVA 2017; Barrientos and Soria 2017). Singapore has externalized the production of high polluting and energy-intensive products including meat and pork, as well as land-intensive cereals and grains. But, Singaporeans continue to consume these foods so the market drives the import of these products. But if Singapore does not or cannot grow or raise many of these items domestically, where does it come from?

A substantial gap in food demand is met by importing from Malaysia (43%), China (29%), India (5%), Indonesia (4%), Australia (4%), and others (15%) (AVA 2017). Beef and mutton are primarily imported from Australia; chicken and pork are sourced from Malaysia as well as Brazil, Netherlands, USA, and France. But those countries are not only producing food for Singapore, and it is critical to consider potential changes in food demand by their own urban populations—as there is likely to be increasing competition for food sources. Singapore might be an island, but it is intimately linked to the cities that rely on the same rural landscapes to meet their own urban food demand. Table 4.1 shows the total population, land area, population density, per capita in USD, GFSI ranking, and percent of total population that lives in an urban area for Singapore and four important importing countries. In terms of buying power, Singapore ranks high against China, Indonesia, and Malaysia with per capita incomes less than one-fifth as compared to Singaporeans.

4.3.1 *The Hinterlands in Malaysia*

In contrast to Singapore, Malaysia ranks 41/113 on the GFSI (The Economist Group 2017); and per capita income is $9502.60 (USD) (The World Bank 2017b)

Table 4.1 Demographics and food security by county

	Total population[a]	Density (p/km^2)[a]	Land area (km^2)[a]	Per capita (USD)[b]	GFSI rank (out of 113)[c]	Population urban[a] (%)
China	1,409,517,397	150	9,388,211	$8,123.20	45	57
Indonesia	263,991,379	146	1,811,570	$3,570.30	69	53
Malaysia	31,624,264	96	328,550	$9,502.60	41	74
Australia	24,450,561	3	7,682,300	$49,927.80	5	89
Singapore	5,733,551	8155	700	$52,960.70	4	100

[a]United Nations, Department of Economic and Social Affairs, Population Division (2016)
[b]The World Bank (2017b)
[c]The Economist Group (2017)

(Table 4.1). Politically, Singapore was part of Malaysia until independence; and, geographically, Malaysia still represents Singapore's hinterlands, functioning as its agricultural landscape. Linked by a land bridge, it is unsurprising that Singapore sources a majority of its food (particularly eggs and vegetables) from Malaysia. In addition to eggs and vegetables, Singapore imports chicken, cooking oil, duck, fish, fruits, and sugar from Malaysia. Malaysia is one of Singapore's most important import sources; however, bilateral relations between Singapore and Malaysia are not reliably stable. There are persistent disputes over water sources as well as reclamation of land dating back to 1965 independence. But, competition comes from both local and globe pressures.

Global market demand for palm oil has increased rapidly in recent decades.[4] Malaysia is the world's second biggest exporter of palm oil, contributing 39% of global production.[5] As of 2016, a total 5,737,985 ha were planted with oil palm in Malaysia, approximately 17.4% of the country's total land area. According to the World Bank, 23.9% of the total land area in Malaysia is classified as agriculture land, which means three quarters of agricultural land is currently occupied by oil palm. Despite growing environmental concerns, palm oil demand is projected to continue to increase.

Kuala Lumpur, the national capitol of Malaysia, is rated as a global city (the only one in Malaysia). It is one of the fastest growing urban places in Southeast Asia in terms of population and economic development. It has an urban population of approximately 7.0 million; 22.3% of the total population of Malaysia (Table 4.2). City food security is dependent on global market demand for palm oil and higher-earning countries' demand for Malaysia produce. One response to the threat of food insecurity is a small, but growing, grassroots movement promoting local food production in Kuala Lumpur. There are a few organizations practicing and promoting urban agriculture in the form of small community gardens, notably TTDI

[4]Global Industry Analysists, Inc., http://www.strategyr.com/MarketResearch/Palm_Oil_Market_Trends.asp accessed 26 October 2017.

[5]Malaysian Palm Oil Council (MPOC), http://www.mpoc.org.my/Malaysian_Palm_Oil_Industry.aspx accessed 26 October 2017.

Table 4.2 Demographic distribution by city[a]

	Total population	density (p/km^2)	land area (km^2)	Total co population (%)
Jakarta	10,483,000	15,847	662	4
Kuala Lumpur	7,047,000	3141	2243	22
Singapore	5,733,551	8155	700	100
Sydney	4,540,000	367	12,368	19

[a]United Nations, Department of Economic and Social Affairs, Population Division (2016)

The Edible Project, a guerilla garden; Suber Community Gardens, a group working to establish community gardens; and Eats, Shoots & Roots a social enterprise building gardens for community residents. City Farm is a young startup company selling hydroponic systems, with a demonstration farm and offering training courses, and a vision of urban homes functioning as farms. However, unless these small operations scale up dramatically, Kuala Lumpur will continue to face fierce competition for its agricultural lands.

4.3.2 *Jakarta: Indonesia's Megacity*

Immediately to the south of Singapore lies Indonesia, with a population of 264 million distributed across 17,508 islands. Food imports from Indonesia to Singapore include fish, fruits, pork, vegetables, and milk. In contrast to Singapore, Indonesia ranks 69/113 on the GFSI (The Economist Group 2017), and has a per capital income of $3570.30 (USD) (The World Bank 2017b). Currently 31.5% of land is in agriculture (The World Bank 2017a). Indonesia was particularly hard hit during the global economic crisis in 1998 and again in 2007, which led to rising food prices and food insecurity. Although an important food source for Singapore, Indonesia is actually a net importing country. It produces higher-value crops for export and imports lower-value crops for consumption. More on the global market will be discussed in a later section, but suffice it to say, Indonesia represents a low-earning, net-importing country with large, dense urban centers, particularly, the megacity of Jakarta.

Jakarta is the capitol of Indonesia with a population of 10.5 million; it is located on the most populated island of Java. As one of the largest cities in the world, urban food security is particularly challenging. However, urban agriculture is not currently a high priority, evident by a lack of supportive government policies and incentives. Most of urban food demand in Jakarta is met through rural and peri-urban sources. The dynamics of peri-urban agriculture cannot be separated from the policy response to rising food prices caused by the global economic crisis.

In 1998, the governor of Jakarta introduced a policy to utilize vacant lands for agricultural development as a temporary solution to satisfy the food demands

(Purnomohadi 2000; AFP 1998). Learning from the economic crisis, a government regulation about food security was introduced in 2002 to strengthen food production, promote equitable food distribution, and to stabilize food prices. It also keeps reserves which limit domestic price volatility when global markets are less stable (Murphy 2013). When the second crisis occurred in 2007, a new law for spatial planning was established to mandate the development of 'agropolitans,' or agricultural towns in rural areas to facilitate the supply of agricultural products from rural to urban centers. In 2009, a law was enacted to protect agricultural land for food production, although the mechanism through which the land can be protected is still unclear.

Unfortunately, these initiatives have continued to be supported; peri-urban agriculture seems to have been regarded as a safety net during the food crisis, but faces neglect under "normal" conditions (Pribadi and Pauleit 2015). And, the fringe of Jakarta has experienced sprawl as the satellite cities of Bogor, Depok, Tangerang and Bekasi urbanize and consume large areas of prime agricultural land (Fitriani and Sumarminingsih 2015). The current masterplan of *Metropolitan Priority Area* (MPA 2010–2030) favors the growth of Jakarta and its satellite cities; there is pressure to use land for development of high density housing, trade and services, and non-polluting and market-oriented light industry (CMEA 2012). While there is some suggestion of potential land for urban agriculture development, there is no land use policy that explicitly supports urban agriculture. Further complicating matters, there is ambiguity in the categorization of urban agriculture found in the decrees of the *Ministry of Public Works* and the *Ministry of Interior*. According to the *Ministry of Public Works*, urban agriculture is not considered a green open space; in direct contrast, the *Ministry of Interior* considers urban agriculture as green open space (Indraprahasta 2013). At the provincial Jakarta level, in the *Jakarta Spatial Plan Document 2010–2030*, there are several articles demonstrating the potential land for urban agriculture development. However, there is no clear policy on urban agriculture as an important consideration in city masterplan and land use. There is a lack of economic incentives offered by the government especially in Jakarta (Indraprahasta 2013), this despite a study by Adiyoga et al. (2004) that concluded that incentives and disincentives were required to encourage communities and to support urban agriculture development.

As food prices rise, lower income households may lose purchasing power. It is possible that as food insecurity increases, it instigates a bottom-up movement for household kitchen gardens; however, it is difficult to predict. Despite high density development, there is some agriculture practiced within the urban boundary of Jakarta; the vast majority is non-registered, thus outside the formal economy. The majority of informal practitioners are low-income individuals and households producing for self-consumption (Cahya 2016). As Jakarta urbanizes its peri-urban fringes, many farmers face a lack of skills to get a job in the urban sector (Arai 2011); thus, they continue to farm, but often on land where they have lost the security of tenure rights or moving to vacant land without legal rights (as squatters). On the other hand, more advantaged social groups are less likely to be involved in urban agriculture (Cahya 2016). Social and economic benefits of urban agriculture in Jakarta are acknowledged despite lack of top-down initiative.

There has been notable contribution in the private sector for development of urban agriculture through corporate social responsibility (CSR) schemes. For example, PT *Epicentrum* gave permission to use undeveloped land to the local community, coordinated by *Family Welfare and Empowerment Community* (*Pemberdayaan dan Kesejahteraan Keluarga or PKK*) (Indraprahasta 2013). Jakarta Berkebun (English translation: Gardening; offshoot of Indonesia Burkebun) is a non-profit social and environmental urban gardening movement co-founded by an architect/urban designer who designs urban farmsteads. And, Gang Hijau (English translation: Green Alley) started as a grassroots installation of potted food producing plants along residential alleyways that has recently been backed by support from the Jakarta Agency of Food Security to fund additional alley gardens.

4.3.3 Localism in Australia

Australia ranks second on the human development index (HDI) (Jahan 2016) and 5/113 on the GFSI (The Economist Group 2017). It has a population of 24.5 million with two major urban centers: Sydney and Melbourne with 4.5 million and 4.3 million people respectively (United Nations, Department of Economic and Social Affairs, Population Division 2016). Australia represents a high-earning country with per capita income is $49,927.80 (USD) (The World Bank 2017b). Singapore imports beef, mutton, pork and fish; fruits and vegetables; and rice, sugar, and milk from Australia. In contrast to Singapore, Australia has a large land mass and a very low population density of 3 people per km^2. It is a country with a lot of land (52.9% total land in agricultural use) that is forward thinking in securing a sustainable food system through policy and land use planning. A prevailing view of the Australian government is that it is a food secure country in which climate change is significant mainly in relation to aggregate levels of food production and to the changing viability of some major food production areas within the country; there is little or no concern with the resilience of food supply chains to the growing urban populations (Burton et al. 2013).

Australia never has been and never will be a food basket for the world (Staight 2011). Although food exports are estimated to feed 40 million people, they represent less than 3% of global food movements. It may not make a significant impact on feeding global populations, but Australia contributes to global food security through knowledge, training and technology transfer. *«We're very good at farming, we're very, very good at food production. We have better knowledge about how to handle difficult climates, droughts and floods and things like that than almost anyone else in the world, and in a time of climate change that's exactly the kind of knowledge that you need»* (Staight 2011).

In parallel to policy-level response to protecting agricultural land and increasing food security, Sydney has experienced an urban agriculture grass-roots movement

(Mason and Knowd 2010). The shift toward local food production comes out of a cultural movement rather than one of necessity (i.e. quality rather than quantity) (James 2016). In contrast to cities in less developed countries, Sydney's residents are less motivated by meeting gaps in food demand than they are in having control over their food sources in a way that promotes social equity and environmental sustainability.

4.3.4 Renting Land in China

Singapore not only imports food from other countries, but has recently leased farmland in China's Jilin Modern Agriculture Food Zone (Chang 2010). Comprised of 1450 km^2, the entire food zone is twice the size of Singapore, and roughly 200 times larger than Singapore's agricultural land area. In 2008, Jilin's meat output was reported as approximately 3.5 million tonnes; nine times Singaporean demand. It is believed that Jilin Food Zone will help lower Singapore's vulnerability in terms of food security. But is outsourcing land to China a sustainable solution? With a population of 1.4 billion and undergoing urbanization at a rate not seen in other countries, China itself rents land in Africa for agriculture, mining of natural resources, and forest products, spurred by the food crisis in 2008. Current per capita income in China is still quite low at $8123.20 (USD; 2016), but it has nearly quadrupled over the last decade ($2099.23 in 2006) (The World Bank 2017b). It ranks 45/113 on the GFSI and 90 on the HDI (The Economist Group 2017; Jahan 2016). Renting land overseas, also referred to as "land grabbing," is the result of climate change and globalization, which can be defined as the increased integration of national economies through technological advances in the ability to ship more products farther distances in reduced time and legal and regulatory changes that reduced the cost of import-exports (Murphy 2013). It has been coined "land grabbing" due to the nature of these large-scale purchases/leases of land with little or no protection or rights given to those already on the land.

The conditions for renting land overseas has a complicated history; notably, in 1995 the early members of the *World Trade Organization* (WTO) established the *Agreement on Agriculture* (AoA), which ended the exemption of agriculture from the *General Agreement on Tariffs and Trade* (GATT). It was a defining decision that explicitly placed agriculture in the international trade system. The result is complex and far-reaching beyond the vision of the WTO. Some countries, mostly higher-earning, higher-developed, benefited greatly through diversification of food and other land-based resources through imports-exports, leading to growing economies, generating employment, and increasing consumers' purchasing power. For some countries, mostly lower-earning, lower-developed, there was a shift to import-dependence that reduced domestic food production, increased migration from rural areas to urban centers where many remained unemployed or underemployed, and also now lack land to grow their own food.

Land became a commodity that could be bought and sold on the global market. As it became cheaper to move food products long distances, it became increasingly more attractive for countries to import food so that domestic land could be developed into higher value uses. But, as countries turned to cheaper places to import from, there was increasing discomfort that markets could fail to provide enough food to meet domestic demand. Markets are based on supply and demand; when demand increases and supply decreases, prices rise. Only those with high purchasing power will be able to access the reduced supply. This has motivated governments to look for other solutions to meet food import needs: namely purchasing or leasing land overseas. At issue is the discrepancy in power between leasor and leasee (and, particularly, those who live on the land but are not involved in the contract). Furthermore, perhaps more alarming is that arable land and freshwater, on which humans depend for survival, has been commoditized with surprisingly little regulation. The political economy of borrowing cheaper land cannot persist once the cheapest land has been exhausted.

4.3.5 One Island in an Agri-Pelago

It takes little imagination to envision the multiple, diverse, cross-scale risks to Singapore's food security even with the few examples described in this chapter. Singapore is a unique case because of its city-state status, but it is clear that no city can consider itself an island. The global market has caused a fundamental shift in the rural-urban relationship; where cities traditionally relied on their rural hinterlands to provide food and other resources, now rural agricultural lands have become the hinterlands of the highest bidding city. As a high-earning, highly developed country, Singapore is in a position to bid high; however, the status quo is likely to change with rising social, environmental, and political issues faced by urban and rural places.

There is a need to conceptualize the global food system as necessarily linked to local contexts. The linear rural to urban paradigm is a fallacy that should be replaced by a more nuanced conceptualization that takes a socio-ecological approach. The market is driven by demand, but there is power underlying the system: power to change policy, power to purchase locally-grown/raised, power to educate, and power to grow instead of purchase. Food security is not just physical; it is psychological. It is about having the confidence that there will be access to food in the short and long term. On the other hand, it is also about being cognizant of the risks and vulnerabilities despite a perception of security, as in the case of Singapore.

4.4 Conclusion

The government of Singapore has developed a food security road map with strategies that range from increased high-tech, high-yield domestic production and food stocks to food source diversification and reduced waste. As a country that is 100% urban with <1% of agricultural land, it is unlikely to achieve food independence—even with the unrealistic mass conversion of rooftops to farms. In this chapter, we illustrated how it is but one island in an agri-pelago of food trans-boundaries. Malaysia, represents Singapore's overseas hinterlands, but global demand for palm oil, combined with rapidly growing Kuala Lumpur both compete with Singapore for arable Malaysian land. Indonesia, Singapore's neighbor to the south, offers a cheap alternative, but policies have been reactionary when global food prices rise and the local population faces food insecurity. Jakarta is a megacity with more than 10 million mouths to feed—urban agriculture primarily limited to informal, kitchen gardens set up by rural farming migrants on vacant land or small hobby gardens by the middle class. In contrast, Australia represents a high-earning, food secure country with a stable government and economy. Ironically, or rather as a result, it has developed its own food security strategies; Singapore's wealth does not have the same buying power. Finally, buying land in China might create a stable food supply in the short-term, but the lack of transparency in China's government means the future is unsure. One thing that is sure, is that Singapore is not alone. Perhaps is it through that shared insecurity that international policies will have support to protect our most valuable resources: land, water, and livelihoods.

References

Adiyoga, W., Dimyati, A., Soetiarso, T. A., Ameriana, M., & Suherman, R. (2004). Public perception on urban agriculture existence in Jakarta and Bandung. *Journal of Horticulture, 14,* 134–149. (in Indonesian).

Agence France Presse (AFP). (1998). *Jakarta governor says poor can farm city land.* City Farmer, Canada's Office of Urban Agriculture. Accessed 2 September, 2017 http://www.cityfarmer.org/jakarta.html.

Arai, K. (2011). From water buffaloes to motorcycles: The development of large-scale industrial estates and their social-spatial impact on the surrounding villages in Karawang Regency, West Java. *Southeast Asian Studies, 49*(2), 161–191.

AVA. (2017). *Annual statistics.* Agri-Food & Veterinary Authority of Singapore. Accessed 14 September, 2017 https://www.ava.gov.sg/homepage.

Badami, M. G., & Ramankutty, N. (2015). Urban agriculture and food security: A critique based on an assessment of urban land constraints. *Global Food Security, 4*(Supplement C), 8–15.

Barrientos, M., & Soria, C. (2017). *Commodities market: Singapore milled domestic rice consumption by year.* Index Mundi. Accessed 3 October, 2017 http://www.indexmundi.com/.

Benton, J. (2016). *Hydroponics: A practical guide for the soilless grower.* Boca Raton, FL: CRC Press.

Burton, P., Lyons, K., Richards, C., Amati, M., Rose, N., Des Fours, L., et al. (2013). *Urban food security, urban resilience and climate change* (p. 160). Gold Coast: National Climate Change Adaptation Research Facility.

Cahya, D. L. (2016). Analysis of urban agriculture sustainability in metropolitan Jakarta (case study: Urban agriculture in Duri Kosambi). *Procedia—Social and Behavioral Sciences, 227,* 95–100.

Chang, A. (2010). Singapore secures China as future food source. In G. Kleyn, (Ed.). *Strategic analysis paper*. Perth, Australia: Future Directions International. Accessed 16 October, 2014 http://futuredirections.org.au/wp-content/uploads/2011/05/1285301165-FDI%20Strategic%20Analysis%20Paper%20-%2024%20September%202010.pdf.

Coordinating Ministry for Economic Affairs (CMEA). (2012). *Development code B1 from master plan for establishing metropolitan priority area for investment and industry in Jabodetabek area in the Republic of Indonesia*. The Republic of Indonesia. Accessed 30 November, 2017 https://open_jicareport.jica.go.jp/pdf/12083945_01.pdf.

Culzac, N. (2014). Japanese plant experts produce 10,000 lettuce heads a day in LED-lit indoor farm. *The Independent*. Accessed 15 September, 2014 http://www.independent.co.uk/news/science/japanese-plant-experts-produce-10000-lettuces-a-day-in-led-lit-indoor-farm-9601844.html.

Food and Agriculture Organization of the United Nations (FAO). (2017). *Food and Agriculture Organization of the United Nations*. Rome: United Nations. Accessed 17 October, 2017 http://www.fao.org/.

Fitriani, R., & Sumarminingsih, E. (2015). Determination of spatial extent of land use in the fringe of Jakarta metropolitan: A semivariogram analysis. *Theoretical and Empirical Researches in Urban Management, 10*(1), 43–54.

Goodman, E. R. (2011). *Aquaponics: Community and economic development*. Diss: Massachusetts Institute of Technology.

Indraprahasta, G. S. (2013). The potential of urban agriculture development in Jakarta. *Procedia Environmental Sciences, 17,* 11–19.

Jahan, S. (2016). *Human development report 2016: Human development for everyone*. New York, NY: The United Nations Development Programme.

James, S. (2016). Sustaining Sydney. In S. James (Ed.), *Farming on the fringe: Peri-urban agriculture, cultural diversity and sustainability in Sydney* (pp. 145–181). Cham: Springer.

Leitch Lepoer, B. (1989). *Singapore: A country study*. Washington: GPO for the Library of Congress.

Ludher, E. (2015). *A case study of Singapore's smart governance of food*. Singapore: Centre for Liveable Cities Singapore.

Manibo, M. (2013). *Panasonic ventures into farming in Singapore*. Accessed 15 September, 2014 http://www.eco-business.com/news/panasonic-ventures-into-farming-singapore/.

Mason, D., & Knowd, I. (2010). The emergence of urban agriculture: Sydney, Australia. *International Journal of Agricultural Sustainability, 8*(1/2), 62–71.

Mickens, M. A. (2012). *Comparative study of lettuce and radish grown under red and blue light-emitting diodes (LEDs) and white fluorescent lamps*. Accessed 15 September, 2014 https://ntrs.nasa.gov/search.jsp?R=2012001573.

Murphy, S. (2013). *Land grabs and fragile food systems: The role of globalization*. Washington, D.C.: The Institute for Agriculture and Trade Policy.

Pribadi, D. O., & Pauleit, S. (2015). The dynamics of peri-urban agriculture during rapid urbanization of Jabodetabek Metropolitan area. *Land Use Policy, 48,* 13–24.

Purnomohadi, N. (2000). Jakarta: Urban agriculture as an alternative strategy to face the economic crisis. In N. Bakker, M. Dubbeling, S. Gündel, U. Sabel-Koshella, & H. de Zeeuw (Eds.), *Growing cities, growing food. Urban agriculture on the policy agenda* (pp. 453–465). Feldafing, Germany: Zentralstelle für Ernährung und Landwirtschaft (ZEL).

Resh, H. M. (2002). *Hydroponic food production: A definitive guidebook for the advanced home gardener and the commercial hydroponic grower*. Boca Raton, FL: CRC Press.

Sabzalian, M. R., Heydarizadeh, P., Zahedi, M., Boroomand, A., Agharokh, M., Sahba, M. R., et al. (2014). High performance of vegetables, flowers, and medicinal plants in a red-blue LED incubator for indoor plant production. *Agronomy for Sustainable Development, 34*(4), 879–886.

Shim, K. F. (1987). Part IV country papers (Contd)—Singapore. In Asian Productivity Organization (Ed.). *Animal feed resources in Asia and the Pacific*. Tokyo: Asian Productivity Organization. Accessed 21 October, 2014 http://www.fao.org/docrep/004/AC153E/AC153E13.htm#ch4.11.

Singapore Tourism Board. (2016). *Tourism sector performance—Q4 2016 report*. Singapore: Singapore Tourism Board.

Staight, K. (2011). *The future of food. Landline, Ultimo*. Ultimo: Australian Broadcasting Corporation broadcast on Jun 26, 2011.

Swee-Hock, S. (2012). *The population of Singapore* (3rd ed.). Singapore: ISEAS Publishing.

Tan, J. (2013). *How much food can Singapore produce?* Department of Biological Sciences, National University of Singapore.

Teng, P. & Escaler, M. (2010). *The case for urban food security: A Singapore perspective*. NTS Perspectives, No. 4, Singapore: RSIS Centre for Non-Traditional Security (NTS) Studies.

The Economist Group. (2017). *Global food security index*. The Economist Group. Accessed 20 October, 2017 http://foodsecurityindex.eiu.com/Country.

The World Bank (2017a). *Agriculture land (% of land area)*. The World Bank. Accessed 26 October, 2017 https://data.worldbank.org/indicator/AG.LND.AGRI.ZS.

The World Bank (2017b). *GPD per capita (current US$)*. The World Bank. Accessed 26 October, 2017 https://data.worldbank.org/indicator/NY.GDP.PCAP.CD.

United Nations, Department of Economic and Social Affairs, Population Division. (2016). *The World's cities in 2016—Data booklet*. (ST/ESA/SER.A/392).

Wang, B. (2012). *4 Story tall vertical farm that is up to ten times more productive than regular farms*. Accessed 15 January, 2016 http://nextbigfuture.com/2012/11/4-story-tallvertical-farm-that-is-up.html.

Warren, E., Hawkesworth, S., & Knai, C. (2015). Investigating the association between urban agriculture and food security, dietary diversity, and nutritional status: A systematic literature review. *Food Policy, 53,* 54–66.

Wilson, G., (2005). *Food from the roof*. Accessed 1 September, 2014 http://www.greenroofs.com/archives/gf_nov-dec05.htm.

Wong, R. (2011). *Solar potential of HDB blocks in Singapore*. Singapore: Energy Studies Institute, National University of Singapore.

Worldometers.info. (2017). Retrieved from http://www.worldometers.info/world-population/population-by-country/. Accessed 27 October 2017.

Jessica Ann Diehl, Ph.D., is Assistant Professor in the Department of Architecture at the National University of Singapore. She completed her Ph.D. in Health and Behaviorial Sciences at the University of Colorado Denver, USA where she was a National Science Foundation (NSF) IGERT Ph.D. Fellow in Sustainable Urban Infrastructure Systems. She was a 2013–2014 Fulbright-Nehru Research Fellow affiliated with the School of Human Ecology at Ambedkar University Delhi, India. She holds a BLA/MLA in landscape architecture from The Pennsylvania State University, and a BA in English with a concentration in creative writing from West Virginia Wesleyan College.

Ching Sian Sia is a Ph.D. Candidate in the Department of Architecture at the National University of Singapore. She completed her Master of Science in Integrated Sustainable Design at the National University of Singapore. She holds a Postgraduate Diploma in Urban Design and a Bachelor in Urban Planning and Development from the University of Melbourne, Australia. In her professional life as an urban planner and designer, she has also completed master planning projects in China, Maldives and Vietnam.

Amanda Jennifer Chandra is currently pursuing a concurrent degree in Bachelor of Arts in Architecture and Master of Landscape Architecture at the National University of Singapore. She is expected to graduate in 2018. In addition to English, she is proficient in both Indonesian and Malay language.

Part II
Governing the Agro-urban Spaces in Europe

Chapter 5
Contraposition, Juxtaposition, and Transposition of the Urban and the Rural

Marco Santangelo

5.1 The Urban and the Rural

To try to understand the relationship between the urban and the rural is a way to understand the socio-spatial organisation of the space we live in. There is, in fact, a certain degree of difficulty in characterizing a space as (completely) urban or rural, so that it would be easier to consider such definitions as things of the past or to try to figure out if, and in which way, a distinction between urban and rural is still relevant.

In the first case, i.e. that a separation between what is urban and what is rural has nowadays ceased to be useful (since when and if for every part of the planet is of course to be investigated[1]), we should then find what socio-spatial organisation has substituted the previous one. The current debate on the characteristics of the urban age may be helpful here (often promoted by reports of international organisation, such as the UN or the *World Bank,*[2] that reminds us that an ever increasing number of the planet population is going to live in urban areas; UN-DESA-PD 2016). The "rural" here is often a part of the discussion in terms of former counterpart to the urban.

---

[1]The perspective that is assumed here is that of the urban-rural relationship in the so-called *Global North*, and the use of this reference is purely for the sake of immediate understanding of a North American and Western Europe geographical reference of urbanisation processes. This perspective does imply, of course, that specificities in such diverse contexts should anyway be taken into account, as well as attention should be paid to the fact that certain phenomena with the same name (e.g. suburbanisation) have occurred in different moments and with different features around the globe.
[2]http://www.worldbank.org/en/topic/urbandevelopment (last accessed: 4 December 2017).

M. Santangelo (✉)
DIST, Interuniversity Department of Regional and Urban Studies and Planning,
Politecnico and Università di Torino, Turin, Italy
e-mail: marco.santangelo@polito.it

E. Gottero (ed.), *Agrourbanism*, GeoJournal Library 124,
https://doi.org/10.1007/978-3-319-95576-6_5

In the second case, that is to find out if the reference to a rural space as different from the urban one still makes sense and how, we may enter the realm of definitions of rural areas: for instance according to the low density of population or the typology of economic activities that are implemented there (see the definition of "first rural" in Bell 2007).

In both cases, anyway, the rural is actually analysed as in contraposition to the urban, so that the relationship between urban and rural spaces can be, essentially, considered as a product of (i) *the aim of defining what is the urban*, and (ii) *of the recognition of the related nature of one typology of space to the other*. This interpretation, which does define in itself neither the urban nor the rural, is rather a hypothesis that helps us better understand both spaces, how they are conceived, and for what reason(s) they are conceptualized.

The use of the term "product", in fact, is a way to highlight that "urban" and "rural" do not exist in Nature as such, they do not happen without human beings having a role in their constitution and, therefore, they are socially and materially produced (thereby referring to Lefebvre's social production of space 1974). It is quite important to stress this point, since there must be no misunderstanding in considering the rural as some sort of idyllic natural environment, pre-human or ecologically sustainable per se, as well as the urban as the realm of artificiality, technology, post-human, unsustainable per se. Much of these misunderstandings, in reality, result from a long history of contraposition of the urban and the rural, a dichotomous vision of a space for frequent, dense interactions among "citizens", and another for slower, sparse, contacts among farmers.[3] Citizens and farmers, and their spaces, have mostly been considered as two different objects of study: in geography, economy, sociology, for instance, there is a distinction between urban and rural aspects of the relevant discipline and, what is more interesting, this is not just a distinction based on the characteristics of the analysed object. Rather, there seem to be a separation between something that is central for human beings development (the urban) and something that is necessary for the same process (the rural). However, necessity is not centrality. This contraposition, in fact, is mostly due to the need of defining what is urban: that a rural area is somehow "there" and providing food and other goods to an urban area is a given, while less obvious are the characteristics of the urban object, especially since it has changed dramatically in the past centuries and seem to host the majority of inhabitants, innovations, institutions, and everything that has to do with development and change.

What is discussed, here, is the artificial nature of such contraposition (Champion and Hugo 2004; Halfacree 2004; Bell 2007). To understand that the dichotomy between urban and rural is a way to better understand the urban is not to falsify this logic. Rather is to make it explicit: urban and rural are different, they have features that distinguish in many ways one from the other, but this does not mean that they

[3]Wirth, in "Urbanism as a way of life" (1938), describes, *«urban-industrial and rural-folk society as ideal types of community, [so that] we may obtain a perspective for the analysis of the basic models of human association as they appear in contemporary civilization»* (p. 3).

are separate or dissociable. Their contraposition can only work as long as it is helpful to analyse one and the other, but it is quite difficult to think that such strongly related spaces can—somehow—be conceived and exist separately, without the other being—also physically—there.

The nature of their relatedness (and the term nature here is intended as the sum of their characteristics) is, therefore, another point that needs attention. To be related does not only allow the urban to be better defined (ultimately as the non-rural), but brings us to search for elements of this relationship that substantiate the idea that there cannot be urban without rural and vice versa. Large part of the scientific debate on urbanisation, however and especially since attention has been given to the explosion of cities since the suburbanisation processes in North America between the end of the 19th and the beginning of the 20th century, has focused on the urban as a pervasive spatial phenomenon, progressively occupying former rural areas with or—in urban sprawl processes—without contiguity. In many ways, attention has been paid to the urban phenomenon because of its sheer dimension and relevance, thus leaving rural areas in the background as functional space for future expansions. The rural may have influenced the image of future urban expansions, as in the case of new urbanisation that brings together the desire for greener, wider, and healthier spaces (from Howard's 1898 Garden City to the single-family or detached house typology that set the standard for suburbanisation and urban sprawl) but, in turn, it needed to be emptied of all the farming activities since the allowed green was that of gardens and parks rather than fields. Referring to green, wider, healthier space leads to another issue that is crucial for the understanding of the relationship between urban and rural, and that transcends the immediate reference to urban and rural as morphological spaces: lifestyles. How life could be conducted in such spaces has been, in fact, considered as different, and dramatically so until the most recent turns in urbanisation processes (Brenner and Schmid 2014). A typical example of the relationship between urban and rural lifestyles would have been described either as the rural area inhabitants as providers of goods to the urban ones or as the rural areas loosing population in favour of fast-growing urban areas (where lifestyle would, of course, be completely different). This is quite an unbalanced vision that sees the rural as resources provider to the urban. Another description would consider the rural as expansion area for the city new neighbourhoods (e.g. with suburbanisation processes in North America and, later, with urban sprawl in Europe). In any case, there is again a contraposition, because where there is the urban lifestyle there cannot be the rural one, and if it still persists it is a matter of time before it disappears. In lifestyles, more than in spaces, the contraposition has been one between progress and lack of progress (as in Wirth 1938). In the following paragraph the contraposition between urban and rural will be analysed taking into account urban phenomena that have actually confused the distinction between one space and the other and have led, ultimately, to think of urban-rural continuum or of a juxtaposition of such spaces.

5.2 Blurring Boundaries: Urban-Rural Morphologies

The iconic images of urban growth produced back in 1925 by representatives of the Chicago School were detailing an urbanisation process set in waves departing from the city's CBD (the Loop) (Parks et al. 1925). What was urbanised was, basically, a blank space waiting for something new to happen and the idea that was given was of a countryside made up of areas to be progressively included in an urbanised continuum. This was not much of a change from what Engels was writing in 1845, about the condition of the working class in England: in his work rural areas were providing workers for cities but, clearly, the focus of the attention was on the latters rather than on the countryside.

The huge impact that research and debates on cities were having since the Industrial Revolution, because of cities fastened pace of change, has certainly favoured the attention towards the growth of urban areas rather than to the concurrent transformation of the rural ones, even if the transformation of traditional agriculture into an agro-industry was as astonishing as other contemporary phenomena. Cities, however, have been considered as the main spatial object to look at, so that specialists either undertook analyses of changes in rural areas[4] or, rather, the rural was one of the components of the urban growth.

What has emerged, from the School of Chicago and from many other interpretations of urbanisation processes (especially in the first half of the 20th century), is that the rural is a space for urban growth rather than something different. The functional and cultural status of the rural space is then related to the possibility of that space to be transformed, sooner or later, in an urban one. If there is not that possibility, then there is no need to define a relation between such spaces, apart from the necessary references to food production and few more.

This cross-eyed look at rural areas has dominated the debate on spatial configurations of human activities on the planet: from Geddes and his definition of conurbation (1915), to Berry and his analysis on counter-urbanisation (1976), to Garreau and his definition of "unstructured open-space environments" in edge cities (1991, p. 459), what was not urban was on hold until an urban function was foreseen, designed and implemented for that space.

This interpretation is, of course, over dramatizing decades of literature for the purpose of showing that (i) *rural areas and functions have been considered residual for many years* and (ii) *that the idea of contraposition does not actually interpret the relationship between urban and rural in the 20th century of mature suburbanisation and increasing urban sprawl*. If, in fact, contraposition between urban and rural areas is possible, both spaces need to have clear and discernible characteristics. When lines are blurred, mainly because the rural is considered as a

[4]Rural geographers, sociologists, economists, among others, that have paid attention to rural issues often in terms of cultural oriented approaches, landscape preservation and specificities that, once again, may signal a contraposition towards urban areas, at least to clarify the object of research (see, for instance, Champion and Hugo 2004; Woods 2009).

space that will soon be occupied by "something" urban, then we can discuss of juxtaposition rather than contraposition.

Gottmann (1961) and Doxiadis (1967) both wrote of the inevitable city of the future, either as megalopolis or representing an urban area that covers most of the globe, in which what is not urbanised is a reserve for natural environments or areas for food production. The recent debate on planetary urbanisation (since the seminal contribution by Brenner and Schmid 2011) does reflect the need to rethink of an urban-rural relationship but, again, it is seen from the point of view of the urban and it juxtaposes onto the rural even if *«[t]he demarcations separating urban, suburban, and rural zones were recognised to shift historically, […] the spaces themselves were assumed to remain discreet, distinct, and universal. While paradigmatic disagreements have raged regarding the precise nature of the city and the urban, the entire field has long presupposed the existence of a relatively stable, putatively "non-urban" realm as a 'constitutive outside' for its epistemological and empirical operations»* (Brenner and Schmid 2011, p. 11). Brenner and Schmid, in 2015, continue to reflect on urbanisation and, therefore, on rural spaces by recognizing that the new geographies of uneven spatial development *«cannot capture adequately through areal models, with their typological differentiation of space between urban/rural»* (Brenner and Schmid 2015, p. 152). What is interesting to highlight, however, is that the debate around the new urban age do not consider urbanisation processes, and thus their rural counterpart, as happening everywhere in the same way: in some parts of the world, e.g. South America, Africa, parts of Asia, rural areas and non-urban zones still play a relevant social, political, and economic role. That this role is intrinsically connected with urban areas is also indisputable. The debate on the urban age and its relevance for the urban/rural debate is critically analysed by Brenner and Schmid also to better understand the contemporary global urban condition (2014). This is, somehow, another signal that much of the reasoning on the rural happens because what has to be better defined is the urban.[5]

5.3 Looking at Practices of Transposition

Recent attention to agriculture and food production, and related cultural values of such practices in blurred spaces, strengthen the relationship between urban and rural areas. As we have tried to show, we have moved from a contraposition of urban and rural spaces (and meanings), to a juxtaposition of the urban and the rural. What we

[5]An interesting example of the constant need to understand what is the urban can be found in the reconsideration of suburban areas development in the US, and of Garreau's edge cities among them (Phelps 2015). Suburban features and mechanisms are still relevant to understand current urbanisation processes and, in this updated debate on suburban morphology, the "rural" is just a background or, as Phelps et al. (2015) wrote *«rural settlements […] are nevertheless in the orbit of urban and suburban centres»* (p. 512).

are witnessing now is, rather, a transposition of the rural into the urban. This does not mean that the unbalanced relationship between an ever expanding urban and a soon to be occupied rural is questioned: such continuing juxtaposition is still relevant and the whole debate on soil consumption is a proof of that phenomenon.[6] It is interesting to highlight, however, practices that are more or less recognizable as "rural" and that are characterising much of the recent rediscovery of a greener, sustainable, healthier lifestyle in contemporary cities (a trio that reminds us of the greener, wider, healthier spaces we used to define, for instance, Howard's Garden City).

It must be recognized that "rural" activities have always been practiced within cities: allotments or community gardens have been both social experiments and a reclaim of a closer contact with more "natural" practices.[7] The history of *Green Guerrillas* and the *Community Garden Movement* in New York in the 1970s[8] is just one of the examples of how green areas with a rural character—i.e. not just park areas for leisure and sport—have been considered as a way to reuse vacant lots, to fight gentrification, to nourish a community spirit in neglected parts of the city.

That story has evolved in many different ways and places: it is quite frequent to see in contemporary urban regeneration projects that a space for urban agriculture is foreseen in masterplans (e.g. in *Berlin's Tempelhof airport gardens*), its temporary nature also allowing to intervene in empty and disused spaces while investors are convinced of the attractiveness of the area. It is also common that former illegally occupied and cultivated plots are formalized in allotments, thus officially entering the list of planned activities in cities (as happened in the above mentioned case of Berlin, as well in many other cities: in Torino, for instance, it is worth mentioning the *Miraorti* experience[9]).

Beside such mixed practices of gardening and agriculture, food production has officially entered the urban sphere with all the innovative practice of hydroponics and soil-less, vertical production of crops (as in the case of *GrowUp Urban Farms* in the United Kingdom). In this case we are not dealing with some residual cultivated field that has survived within the sprawling city, or some rare vineyard within a royal park: we refer to new entrepreneurs that decide to grow crops in the city and to do it notwithstanding, or because, they are doing it within the urban area (e.g. the *Edible Gardens* start up in Singapore). Many more examples are possible, but to provide a list of these multifaceted practices is not our goal: what is important of

[6]See, for instance, reports and documents of the European Environment Agency (https://www.eea.europa.eu/themes/soil) (last accessed: 4 December 2017) or of the Food and Agriculture Organization of the United Nations (http://www.fao.org/land-water/en/) (last accessed: 4 December 2017).

[7]We could also refer to victory or war gardens: that is a very specific experience, limited in space and time, although it is often referred to when dealing with nowadays community gardens.

[8]https://www.nycgovparks.org/about/history/community-gardens/movement (last accessed: 4 December 2017).

[9]http://miraorti.com (the website is in Italian only; last accessed: 4 December 2017).

these trends is that they happen within a combined framework of urban and rural characteristics.

In many of those practices, the role of sociability, associated with community building, is crucial and is related with the possibility that a less hectic lifestyle, in an enjoyable green space, could improve the living conditions of urban dwellers. It is, somehow, a revised version of Wirth's urbanism. On the other hand, the transposition of something rural in the urban refers to the updating of the food producing function within cities but without high soil consumption. These are, of course, very different understandings of rurality within the urban but what seems to keep them together is a cultural approach to the rural, an approach that search to combine two apparently conflicting spheres.

5.4 Conclusions: Still in Need of Knowing What Is Rural (?)

The brief analysis that has been proposed in the previous paragraphs had the main objective of understanding better what is the urban through the rural (in the *Global North*). A secondary effect of such objective could certainly be that we may end having a clearer idea of what the rural is but, as we have seen, either we assume that we can interpret urbanisation processes as if the *Industrial Revolution* has not happened, or we have to deal with the complexity of a still to be fully understood planetary urbanisation.

Even the not so comforting idea that there is a blurred urbanisation, in which we can find both urban and rural characteristics, does not help in making progress, mainly because it points to scenarios of infinite city and megalopolis. In such perspectives, the rural is pragmatically just the space for growing crops for urban dwellers, emptied of any life (and lifestyle) of its own. This is, of course, a catastrophic vision of what we have called as juxtaposition of the urban and the rural: a relationship that still sees each of the two "objects" with specific characteristics. In many ways, if we choose to look at the urban-rural relationship from a morphological point of view, the juxtaposition is still a relevant idea to understand what surrounds us. Suburbanisation (even in its most recent shapes) and urban sprawl are spatial phenomena of strong interconnection of elements of urban and rural areas (e.g. population and building density, economic activities, infrastructures, ecosystem services, etc.).

There are certain phenomena, however, that needs to be included in this picture and that cannot be explained only in terms of juxtaposition. Urban agriculture, community gardening, formalisation of former illegal allotments, urban food policies, food security, etc. are all examples of practices and practice-related policies that are configuring new urban spaces with what could also be described as rural activities. That is why we have introduced the idea of transposition of the rural onto the urban: not just as the continuation of rural practices in spaces that have become urban, but as newly established activities within the city.

The coexistence of juxtaposed and transposed practices is, probably, the contemporary shape of the relationship between the urban and the rural and, even if this may seem to not clarify what is urban and what is rural, it actually helps us to better figure out new possibilities for urban areas as well as a future for the rural ones.

References

Bell, M. M. (2007). The two-ness of rural life and the ends of rural scholarship. *Journal of Rural Studies, 23,* 402–415.

Berry, B. (1976). *Urbanization and counterurbanization*. Beverly Hills (CA): Sage.

Brenner, N., & Schmid, C. (2011). Planetary urbanisation. In M. Gandy (Ed.), *Urban constellations* (pp. 10–13). Berlin: Jovis.

Brenner, N., & Schmid, C. (2014). The 'urban age' in question. *International Journal or Urban and Regional Research, 38*(3), 731–755.

Brenner, N., & Schmid, C. (2015). Towards a new epistemology of the urban? *City, 19*(2–3), 151–182.

Champion, T., & Hugo, G. (Eds.). (2004). *New forms of urbanization: Beyond the urban-rural dichotomy*. Aldershot: Ashgate.

Doxiadis, C. A. (1967). *Ecumenopolis: The inevitable city of the future.* Athens: Athens Center of Ekistics.

Garreau, J. (1991). *Edge city: Life on the new frontier*. New York: Anchor Books.

Geddes, P. (1915). *Cities in evolution: An introduction to the town planning movement and to the study of civics*. London: Williams & Norgate.

Gottmann, J. (1961). *Megalopolis: The urbanized northeastern seaboard of the United States*, New York: The Twentieth Century Fund.

Halfacree, K. H. (2004). Rethinking 'rurality'. In T. Champion & G. Hugo (Eds.), *New forms of urbanization: Beyond the urban-rural dichotomy* (pp. 285–304). Aldershot: Ashgate.

Lefebvre, H. (1974). *La production de l'Espace*. Paris: Anthropos.

Park, R. E., Burgess, E. W., & McKenzie, R. D. (1925). *The City*. Chicago and London: The University of Chicago Press.

Phelps, N. A. (2015). *Sequel to suburbia: Glimpses of America's post-suburban future*. Cambridge (MA): The MIT Press.

Phelps, N. A., Tarazona Vento, A., & Roitman, S. (2015). The suburban question: grassroots politics and place making in spanish suburbs. *Environment and Planning C: Government and Policy, 33,* 512–532.

UN-DESA-PD (United Nations Department of Economic and Social Affairs, Population Division). (2016). Policies on spatial distribution and urbanization: Data booklet, (ST/ESA/ SER.A/394). Retrieved December 4, 2017.

Wirth, L. (1938). Urbanism as a way of life. *The American Journal of Sociology, 44*(1), 1–24.

Woods, M. (2009). Rural geography: Blurring boundaries and making connections. *Progress in Human Geography, 33*(6), 849–858.

Marco Santangelo is associate professor of Geography at the DIST department (Interuniversity Department of Urban and Regional Planning and Studies) of the Politecnico di Torino. His research interests are in the field of urban and regional geography, paying specific attention to spatial development models and practices and to the role of digitalization of contemporary life in the transformation of urban environments. He has worked in national and international research projects, the most recent being: an ESPON on regional strategies for sustainable and inclusive development (a two year European Union funded project); an international collaboration with the National University of Singapore on socio-ecological assessment of urban green areas (one year); the analysis of potentials and limits of the smart city for the Torino city area (funded by a bank foundation for 3 years). He is currently responsible for the internationalisation strategies of the DIST department and for the ERASMUS + initiatives of students and staff international mobility for the courses in planning (bachelor and master of science levels). He is also member of the board of the Ph.D. programme in Urban and Regional Development.

Chapter 6
Towards and Effective Implementation of Green Infrastructure in Rural Areas. Challenges and Options for a Substantial Integration with Spatial Planning

Carlo Rega

6.1 Introduction

The concept of *Green Infrastructure* (GI) has been gaining popularity over the last years among scholars, conservationists and planners, especially in Europe and in the US (Garmendia et al. 2016). The European Commission launched GI as a key element of its *Biodiversity Strategy to 2020* (EC 2011) and in 2013 issued a specific communication to the Parliament and the Council in which GI is defined as a «*strategically planned network of natural and semi-natural areas with other environmental features designed and managed to deliver a wide range of ecosystem services (ES)*» (EC 2013, p. 3).

The aims of GI as established by the EC are manifold and include conservation of ecologically valuable areas, contrasting climate change, provision of multiple ES, but also broader socio-economic objectives such as building a greater sense of community, combating social exclusion, strengthening regional and urban development, and creating jobs.

The definition of GI as proposed by the EC is widely cited, but it is not uncontroversial. Some authors have argued that such a wide array of objectives and functions attached to GI may be a double-edged sword. On the one hand the concept of GI can act as a 'boundary object', i.e. a concept or idea robust enough to enable cross-communication but broad and plastic enough to be interpreted

C. Rega (✉)
ITERAS—Research Centre for Sustainability and Territorial Innovation,
Via C. Colombo, 70126 Bari, BA, Italy
e-mail: carlo.rega@ec.europa.eu

Present Address:
C. Rega
European Commission—Joint Research Centre, Directorate D—Sustainable
Resources, Via E. Fermi 2749, 21027 Ispra, VA, Italy

E. Gottero (ed.), *Agrourbanism*, GeoJournal Library 124,
https://doi.org/10.1007/978-3-319-95576-6_6

differently across communities and stakeholders (Star and Griesemer 1989, as cited in Garmendia et al. 2016). Potentially, the GI concept can thus provide a common frame in which planners, conservation biologists and policy-makers can act together towards a common aim. On the other hand, there is a risk that such a broad definition obscures the inherently conflicting nature of different objectives or lead to the identification of any "green space" as part of GI (Garmendia et al. 2016). It is also argued that the way the GI concept is framed conveys the neoliberal and anthropocentric paradigm whereby ecosystems are considered as "natural capital", in the same way as a physical or financial assets that shall be preserved insofar as it delivers "services" and benefits for humans (ibid).

Nonetheless, the EC *Biodiversity Strategy* and GI as one of its main tools undoubtedly represent a major policy boost in Europe for the development of nature-based solutions and the incorporation of ecological rationality in decision-making and, in particular, spatial planning. As maintained by Snäll et al. (2016), the GI is in fact an inherently spatial concept that should involve bringing together spatially explicit data and scientific spatial modeling and planning methods.

As several authors have highlighted (e.g. Liquete et al. 2015), the definition of GI by the EC encompasses three distinct—though interlinked—concepts: (1) the *planning* and management component; (2) *multifunctionality*, in terms of provision of multiple ES, and (3) *connectivity* as an essential functional characteristics of GI. All these aspects are relevant for the effective implementation of GI and, at the same time, pose challenges that shall be recognized and addressed. In the remainder of this chapter, we discuss the implications and challenges posed by the consideration of these three dimensions for spatial and land use planning, and describe concepts and approaches that can be adopted when planning for GI, with a focus on rural areas.

6.2 Green Infrastructure in Europe: Concepts and Objectives

6.2.1 The Planning Component

The planning component is the first element of the GI. The need to integrate the conservation objectives of the EU into spatial planning for their effective implementation was already put forward in the *Biodiversity Strategy to 2020*, which under Sect. 6.4.1 states that «*The Commission will further encourage collaboration between researchers and other stakeholders involved in spatial planning and land use management in implementing biodiversity strategies at all levels, ensuring coherence with relevant recommendations set out in the European Territorial Agenda*» (EC 2011, p. 8).

GI is not envisaged merely as a set of already existing areas with some ecological value that shall be safeguarded. Its definition as a "*strategically planned network*" calls for human agency at work to *design* such a network. The concept of a network of green areas may not sound as anything new to spatial planners. The idea of *ecological networks* made up by core areas, buffer zones and corridors connecting them has preceded the GI definition and has been extensively used by ecologists and landscape planners. The *Natura 2000* network represents an outstanding example of this at European scale, and planners have been dealing with *Natura 2000* sites at local scales for years now. However, the main rationale behind the establishment of *Natura 2000* and ecological networks at various scale was the conservation of biodiversity and natural values, and preservation from (mainly anthropic) pressures. The GI definition entails a more complex conceptualization, with its emphasis on strategy and delivery of multiple ES.

Firstly, strategic planning requires a strategy, which means the identification and selection of a set of objectives that can and should be pursued through GI. This implies acknowledging that win-win solutions are not always possible and trade-off between different and conflicting goals shall be clearly identified. Typically, trade-offs arise between provisioning and regulating/cultural services, but also within the same class of services. Even the link between biodiversity conservation and ES shall be not taken for granted. It is acknowledged that biodiversity underpins most of the ecosystem functions and processes that can translate into benefits for people (see e.g. Maes et al. 2012), hence, conservation policies will likely be beneficial also for ES supply. However, as Garmendia et al. (2016) point out, the link between biodiversity conservation and provision of ES is complex and context-, time- and pattern-dependent. Therefore, not all measures aimed at preserving biodiversity will result in increased provision or multiple ES and this automatic relation shall not be implicitly assumed in the design of GI. This is even truer if broader socio-economic objectives are taken into account: human well-being and biodiversity conservation will often trade-off, and choices must be made (McShane et al. 2011).

Furthermore, even is the focus is on ES only, not all landscape configurations have the same capacity to deliver ES, and there is no optimal landscape structure able to optimize the supply of all services at once, and across scales. So, which objectives and ES to prioritize shall be made explicit when designing GI. After all, identifying trade-offs and make informed choices is what decision-makers and planners should routinely do when defining different land use options, and doing this in the context of GI design should be no exception. However, how to *methodologically* deal with this poses further questions and challenges, discussed in the following subsection.

6.2.2 Multifunctionality in Practice: The Delivery of Multiple Ecosystem Services

GI should be specifically aimed at delivering multiple ES. Therefore, any planning action/measure aimed at preserving, enhancing or creating GI should be assessed against its potential to increase ES supply. Yet, this entails firstly, that planners have at disposal spatially explicit information on the current level of ES supply in the territories affected by the plans; and secondly, that they are able to model changes in ES as a results of the actions foreseen by the plan. There is however a number of recurrent issues with mapping ecosystems service that should be considered to this regard.

A first issue concerns the very conceptualization of ES. In the wide cited "cascade framework" originally proposed by Haines-Young and Potschin (2010) and recently refined by La Notte et al. (2017), biodiversity and ecosystems are linked stepwise to human wellbeing through the flow of ES (Fig. 6.1). Ecosystems are conceptualized as a network of biophysical structures and ecological processes, featuring flows of energy and matter through space and time. Only a subset of the ecosystems' characteristics and properties (i.e. ecosystem functions) have a utility for people that, once realized, turn into a proper service. Available ES maps not always refer to the final service used or consumed by people. In many cases, the maps actually refer to other steps of the cascade, such as the ecosystem function or even some characteristics of the ecosystem structure. Moreover, available indicators/maps might refer to ES flow, stock, potential or demand (Maes 2017). Stocks are generally easier to map than flows, which are dynamic in time, and, particularly for large areas, the potential ES supply is easier to map than the realized service. For planners to understand what is actually mapped by available datasets is thus crucial to make informed choices when planning for the GI.

Scale-related issues and spatial resolution are other widely cited points in ES mapping. Different services can be mapped at very different scales because the underlying ecological process occur at different scales. When considering multiple ES, as it should be done when planning for GI, different available ES maps will typically have different spatial accuracy and mapping resolutions. The harmonization of different datasets for planning purposes requires in this case to resort to spatial operations and adjustments, which might not be straightforward and lead to increased uncertainties and propagation of errors (Maes 2017).

Discussing the several implications on scale-related issues in ES mapping in details goes beyond the scope of this contribution, but it is important to highlight a sometimes neglected aspect relevant for spatial planning. Scale may matter in different senses; to illustrate this, we can consider two ES as examples: carbon storage by trees and pollination. The reduction of CO_2 concentration provided by a forest patch acts at a global scale, as the greenhouse effect is a global issue and the whole atmosphere can be considered a single system to this regard. Conversely, for insects-mediated ES like pollination, the range of influence of service providers can span from few hundred meters to few Km, and the direct service (e.g. increase of crop) is provided locally.

However, the scale at which the landscape structure affects the supply of the service is different; although carbon sequestration is a global service, the capacity of a single land patch to supply it can be considered relatively not affected by the surrounding landscape configuration, and the overall service provided by a certain number of patches can be approximated through a linear sum. For

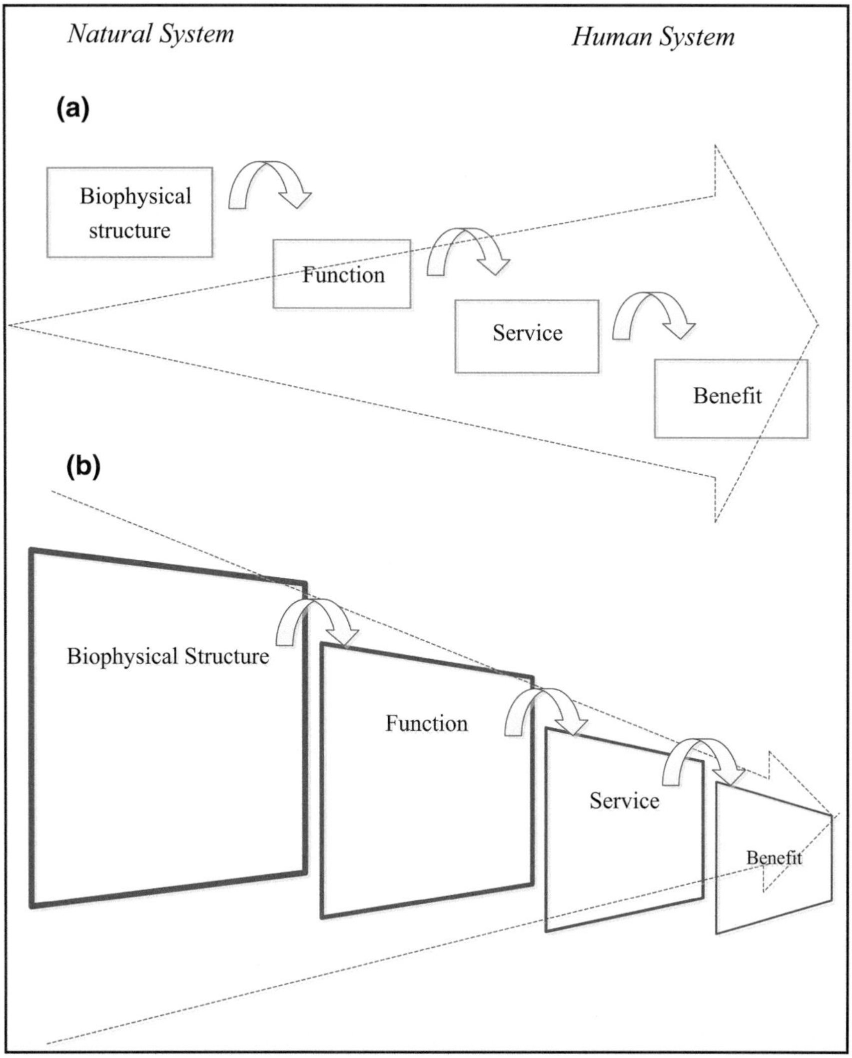

Fig. 6.1 **a** The traditional cascade framework with emphasis on end-use benefits; **b** re-interpretation of the cascade framework, with emphasis on the underpinning complexity of the ecological system. *Source* La Notte et al. (2017)

pollination, landscape structure within the flying range of service providers, in terms of presence and spatial arrangements of different land use types and their vegetation traits, is instead a key factor affecting the pollination supply potential at a certain location (Zulian et al. 2013), whilst only the demand of the service can be considered independent from it. In the first case, the service provided will be mostly dependent only on the area of forest cover in a certain territorial unit, regardless of its spatial arrangement, whilst the opposite would be true for pollination. This has implications on the efforts needed first to model and then to assess changes in ES delivery under different planning scenarios.

This implies that information conveyed by land cover/land use maps traditionally used in spatial planning is often not sufficient for proper ES mapping. Assessing the level of different ES based uniquely on land cover maps has been and still is a widely used method—see the famous work by Burkhard et al. (2009). However, it is increasingly recognized that information on land cover alone cannot capture the complexity of the processes underlying ES delivery. For some ES, the supply capacity of a land cover patch is strongly dependent on the quality or condition of the ecosystem, which is an additional piece of information. Timber production within a forest patch for instance depends on the species composition and age class distributions of the forests (Erhard et al. 2017); similarly, the potential to supply pollination depends on first approximation on the land cover category, but within each category is strongly affected by the presence of flowering vegetation and some key species particularly attractive for bees. Agricultural areas provide another important example, as their capacity to supply (or their reliance upon) ES is strongly dependent not only by the type of cover, but also by farming practices and intensity of management. Not all this information is provided by land cover maps alone.

6.2.3 Connectivity

Connectivity is the third main characteristic of the GI. In ecological terms, connectivity refers to the possibility of dispersion of individuals across space from source habitats patches to destination patches (Kukkala and Moilanen 2017). Connectivity has a structural and a functional component. The first one describes the physical characteristics of a landscape (topography, morphology etc.), whilst the second one refers to how individuals and populations move through the space, depending on their ecological characteristics (Rudnick et al. 2012). In the nomenclature of ecological networks, used also by many authors for GI (e.g. Liquete et al. 2015), structural connectivity is described by identifying core areas, i.e. areas with high ecological value able to support a high degree of biodiversity, and corridors representing the connection between them. Areas with less ecological value than cores, but still important to support biodiversity, are also usually identified (sometimes termed buffer zones or complementary network). Non-habitat areas are usually defined in terms of permeability or resistance, a measure of the difficulty of dispersion for considered organisms. Usually, connectivity at large scales is measured taking into consideration the dispersal ability and habitat requirements of large mammals, but other taxa might well be considered depending on the aim of the analysis. The degree of connectivity of a network of areas, and even the elements that can be considered part of the network, thus strongly depends on the species considered.

As discussed by Kukkala and Moilanen (2017), three different types of connectivity should be considered when dealing with Ecosystem Services. The first one concerns local area size requirements: some ES cannot be provided by small areas (e.g. outdoor recreation) or require sufficiently large areas for the underlying ecological process to operate (e.g. flood protection). In turn, these areas should be part of a regionally connected network to be able to sustain biodiversity and the ecosystem functions that lead to ES delivery. The second type of connectivity refers to the link between the site when the service is generated and the site where beneficiaries use it. This link between supply and demand is usually referred to as ES flow. In some cases, the demand can be far from the supply but still be met, as for instance for food production or outdoor recreation. In these cases in fact either the supply or the demand can displace into space for large distances. In other cases (e.g. pollination or pest control), spatial proximity between suppliers and beneficiaries is required for the flow to exist. The direction of the flow shall be taken into account as well. Pest control and pollination, for example, can be considered as multi-directional services, whilst water flows and related ES are unidirectional. The third type of connectivity regards the degree of dispersion needed to guarantee equitable access to the ES provision across different territorial unites. For instance, drinking water provision or flood protection can be considered realized ecosystem services only if they are benefited by a large number of users (ibid).

All these aspects of connectivity are relevant in a planning perspective and should be addressed in GI design. Importantly, connectivity is a multiscalar concept and shall be addressed at different scales in relation to different ES and planning goals. The connectivity relevant for certain purposes (e.g. conservation of large mammals) may not be at a spatial scale relevant to local land use planning (Rudnick et al. 2012). Finally, spatial planning defines future land uses, so supply and demand of ES GI should be considered not only in relation to the present situation, but also with regard to planning scenarios envisaging the localization of new demand (e.g. new residential developments).

6.3 Planning for Green Infrastructure in Agricultural Areas

One of the most important implications for spatial and land use planning ensuing from the considerations reported in the previous sections is that not all "green spaces" can be considered part of GI, but only those meeting the multifunctionality and connectivity requirements. While this is a general principle that should be applied in all GI mapping activities, specific approaches are needed when planning for GI in different contexts or at different scales. Arguably, how to consider ES provision and connectivity will differ if the focus is on large scale networks aimed at biodiversity conservation at regional level or instead on urban areas. Similarly, the Green Infrastructure in agricultural land will have different characteristics and different methods will be needed to map and plan it. In this section we describe two different approaches taken from recent literature to illustrate the multiscalar nature of GI and to highlight in particular the specificities of GI planning in agrarian landscapes.

A comprehensive approach to map the major element of GI at the European scale is provided by Liquete et al. (2015). These authors propose a spatially-explicit methodology based on: (1) *the quantification of the natural capacity to deliver ecosystem services* and (2) *the identification of core habitats and wildlife corridors for biota*. The first step is achieved by identifying available maps of regulating and maintenance ES and by combining them so to derive an aggregated measure of ES supply at grid level (1 km spatial resolution). The connectivity component was addressed by identifying core areas able to support the presence of large mammals and the major corridors connecting them. The outputs of these two steps were integrated through spatial overlay and the result is the identification of two classes of GI elements: (i) *core GI, comprising the best functioning ecosystems, crucial to maintain both natural life and natural capital*; and (ii) *subsidiary GI, made of other relevant areas sustaining ecosystem services and wildlife*.

The focus of the analysis by Liquete et al. (2015) was on regulating and maintenance service, because most of the provisioning and cultural services are mainly driven by human inputs and do not necessarily enhance natural processes. This leaves as a result the great majority of agricultural land outside the identified core and subsidiary GI. However, besides cultivated crops, agricultural land may include several natural and semi-natural features that can form part of GI, as trees, hedgerows, ditches, ponds, small woodlots, as well as cultural landscape features such as terraces or traditional buildings. Recent researches have shown how the presence of semi-natural vegetation in agricultural land is correlated to the provision of multiple ES, including pollination, pest control, protection from soil erosion and aesthetic appreciation (Smith et al. 2017).

Such areas can therefore be element of the GI and given the importance of agriculture in terms of coverage in Europe (almost half of the whole territory), approaches are needed to measure the actual and potential contribution of agricultural land to GI. European policies explicitly envisage that agriculture contribute to environmental sustainability: target 3 of the *Biodiversity Strategy* states that the CAP should contribute to preserve biodiversity and enhance the provision of ES. The CAP itself promotes, under Pillar 1, the establishment of *Ecological Focus Areas* and, under *Rural Development Programs* (Pillar 2), the establishment and maintenance of natural and semi-natural elements of the agrarian landscape through agri-environmental schemes. Other measures of the *Rural Development Programs* aim at preserving cultural features (e.g. traditional building) and traditional, low-intensive farming practices that are associated to the maintenance of specific habitats with high potential for ES delivery. Semi-natural grasslands with moderate grazing load and other *High Nature Value* farming areas are significant examples of agricultural land able to conserve biodiversity and provide multiple ES (Paracchini et al. 2008).

Semi-natural elements in more intensive agricultural land are typically interspersed in the agrarian matrix and not large enough to enable the movement of large mammals. Structural and functional connectivity in these areas should thus be measured in a different way, by taking into account the organisms that contribute to functional biodiversity in agroecosystems, like pollinators and natural predators providing pest control.

A spatially-explicit approach to map and design GI in medium-high intensive agricultural land was proposed by Estreguil et al. (2016) on a case study area in the arable land of the *Po Plain* in *Southern Lombardy*. In this case, the connectivity analysis took into account the dispersal range of insects providing two key ES in cropped land, pollination and pest control. Agricultural cells potentially belonging to GI (spatial resolution = 100 m) were identified as those with high shares of semi-natural vegetation (woody and herbaceous area >20% of the total area). This threshold is considered in the literature as the level above which biodiversity in cropped areas can be supported (Tscharntke et al. 2011) and can provide ES. The measure of semi-natural vegetation abundance in cropped areas was based on the combination of available detailed spatial layers such as the *Copernicus forest High Resolution layer*[1] and a map of semi natural grassy vegetation abundance in agricultural land developed by García-Feced et al. (2014). The probability of connections between any two cells with high level of semi natural vegetation was determined by considering the dispersal range of European wild bees as reported in literature; the importance of each cell to the total network connectivity was determined by measuring the number of links to other cells.

Simulations were then made by hypothesizing the enhancement of semi natural vegetation in agricultural areas with current low proportion (<20%) to determine the potential achievable increase in connectivity. The cost of the increase of semi natural vegetation was also estimated, by considering the potential yield lost in converting part of the cropped area to habitat to reach the 20% threshold. In this way it was possible to identify the best options for targeting measure to increase the ecological equipment of agricultural areas (i.e. to design new components of GI) under different policy objectives: (1) *maximization of the connectivity without budget limitation*; (2) *maximization of the total connectivity with a given budget*; (3) *achievement of a predetermined level of connectivity with minimum loss of agricultural land*; (4) *achievement of a predetermined level of connectivity with minimum loss of income for farmers*.

These examples illustrate how the general principles of the GI described in Sect. 6.2 can be implemented in different ways when planning for GI in different contexts. In agricultural-dominated landscapes, the usual "core areas" of GI are unlikely to be present. In these cases, analyses at a more detailed spatial resolution should be carried out, to account for those features that, within agroecosystems, can

[1]See: http://land.copernicus.eu/pan-european/high-resolution-layers/forests. Last accessed 12/03/2018.

provide ES and be functionally connected. This requires however detailed spatial information on the occurrence of such features—tree lines, hedgerows, herbaceous areas.

6.4 Conclusions

In this chapter we have discussed three main aspects of the GI and some of their more relevant implications for spatial planning. The GI is a planned network acting at different scales: common to all elements of GI at any scales is the capacity of providing multiple ES and connectivity with the rest of the network. Consideration of ES supply by elements of GI requires planners to handle spatially-explicit information on current and possibly future state of ES stocks and flows in a certain area. However, ES mapping is a large and multidisciplinary domain, spanning from hydrology to ecology, biology and social sciences. All the expertise needed to address relevant ecosystem services in a given area are typically not available within a team of spatial planners. Interaction with sectorial experts and modelers would be ideal in these cases, but this is often costly for planning authorities and time consuming. Ready-to-use tools are therefore needed.

Fortunately, methods for mapping and assessing ecosystem services have been developing steadily over the last years. In Europe, this has happened mainly under the thrust of the specific working group established under the *Common implementation Framework of the Biodiversity Strategy* to support Member States in mapping the state of ecosystems and their services in their national territory. As a result, a wide array of tools and models has have been developed (see Englund et al. 2017; and Palomo et al. 2017, for on overview). Increasingly, suites of models and maps are made available on line and through open access: examples include the InVEST suite (Sharp et al. 2016); ESTIMAP (Zulian et al. 2014, for pan-European maps); and web-based platform such as the ESP Visualization tool (ESP-VT[2]) (Drakou et al. 2015) that allows users to share information on ecosystem services maps, data and mapping methods. Software for connectivity analysis in support of spatial planning are increasingly available too.[3]

Connectivity is in turn a multiscalar concept and shall be addressed in different ways depending on the planning context and the scope of the analysis. In Sect. 6.3 we have provided two examples taken from recent literature of different approaches that can be used to map GI at European level and at a more detailed spatial scale specifically in agricultural land, each dealing with connectivity in a different way, fit to the specific purpose and scale of the study.

---

[2]http://esp-mapping.net/Home/. (Accessed 12/03/2018).

[3]See e.g. Linkage Mapper (http://www.circuitscape.org/linkagemapper) or GUIDOS toolbox http://forest.jrc.ec.europa.eu/download/software/guidos/ (Accessed 12/03/2018).

The systematic consideration of GI principles in spatial planning holds a great potential to reduce habitat loss and fragmentation, enhance the delivery of different ES and contribute to biological conservations. To this end, spatially explicit approaches and methods for mapping GI are required to effectively support the planning process. In turn, planning practice should be flexible enough to incorporate new concepts and methods to explicitly consider both ES and connectivity. Considerations of trade-offs and across space and time between different ES should also be a substantive part of GI planning.

References

Burkhard, B., Kroll, F., Müller, F., & Windhorst, W. (2009). Landscapes' capacities to provide ecosystem services–A concept for land-cover based assessments. *Landscape Online, 15*(1), 22.

Drakou, E. G., Crossman, N. D., Willemen, L., Burkhard, B., Palomo, I., Maes, J., et al. (2015). A visualization and data-sharing tool for ecosystem service maps: Lessons learnt, challenges and the way forward. *Ecosystem Services, 13,* 134–140. https://doi.org/10.1016/j.ecoser.2014.12.002.

Englund, O., Berndes, G., & Cederberg, C. (2017). How to analyse ecosystem services in landscapes-A systematic review. *Ecological Indicators, 73,* 450–492.

Erhard, M., Banko, G., Malak, D. A., & Martin, F. S. (2017). Mapping ecosystem types and conditions. In B. Burkhard, & Maes, J. (Eds.), *Mapping ecosystem services* (pp. 75–80). Sofia: Pensoft Publishers.

Estreguil, C., Caudullo, G., Rega, C., & Paracchini, M. L. (2016). *Enhancing Connectivity, Improving Green Infrastructure.* EUR28142 EN. Luxembourg: Office for Official Publications of the European Union, https://doi.org/10.2788/170924.

European Commission. (2011). *Our life insurance, our natural capital: An EU biodiversity strategy to 2020 European Commission.* COM (2011) 244, Brussels (2011). Accessed 12/03/2018.

European Commission Communication from the Commission to the European Parliament, the Council, the European Economic and Social Committee and the Committee of the Regions. (2013). *Green infrastructure* (gi)—*enhancing Europe's natural capital.* COM (2013) 249 final, Brussels. Accessed 12/03/2018.

García-Feced, C., Weissteiner, C. J., Baraldi, A., Paracchini, M. L., Maes, J., Zulian, G., et al. (2014). Semi-natural vegetation in agricultural land: European map and links to ecosystem service supply. *Agronomy for Sustainable Development, 35*(1), 273–283.

Garmendia, E., Apostolopoulou, E., Adams, W. M., & Bormpoudakis, D. (2016). Biodiversity and green infrastructure in Europe: Boundary object or ecological trap? *Land Use Policy, 56,* 315–319.

Haines-Young, R. H., & Potschin, M. (2010). The links between biodiversity, ecosystem services and human well-being. In D. Raffaelli, & M. Frid (Eds.), *Ecosystem ecology: A new synthesis. BES Ecological Reviews Series* (pp. 110–139). Cambridge: Cambridge University Press.

Kukkala, A. S., & Moilanen, A. (2017). Ecosystem services and connectivity in spatial conservation prioritization. *Landscape Ecology, 32*(1), 5–14. https://doi.org/10.1007/s10980-016-0446-y.

La Notte, A., D'Amato, D., Mäkinen, H., Paracchini, M. L., Liquete, C., Egoh, B., et al. (2017). Ecosystem services classification: A systems ecology perspective of the cascade framework. *Ecological Indicators, 74,* 392–402. https://doi.org/10.1016/j.ecolind.2016.11.030.

Liquete, C., Kleeschulte, S., Dige, G., Maes, J., Grizzetti, B., Olah, B., et al. (2015). Mapping green infrastructure based on ecosystem services and ecological networks: A Pan-European case study. *Environmental Science and Policy, 54,* 268–280.

Maes, J., Egoh, B., Willemen, L., Liquete, C., Vihervaara, P., Schägner, J., et al. (2012). Mapping ecosystem services for policy support and decision making in the European Union. *Ecosystem Services, 1*(1), 31–39.

Maes, J. (2017). Specific challenges of mapping ecosystem services. In B. Burkhard, & J. Maes (Eds.), *Mapping ecosystem services* (pp. 87–89). Sofia: Pensoft Publishers.,

McShane, T. O., Hirsch, P. D., Trung, T. C., Songorwa, A. N., Kinzig, A., Monteferri, B. … Welch-Devine, M. (2011). Hard choices: making trade-offs between biodiversity conservation and human well-being. *Biological Conservation*, *144*(3), 966–972.

Palomo, I., Adamescu, M., Bagstad, K. J., Cazacu, C., Klug, H., & Nedkov, S. (2017). Tools for mapping ecosystem services. In B. Burkhard, & J. Maes (Eds.), *Mapping ecosystem services* (pp. 70–73). Sofia: Pensoft Publishers.

Paracchini, M. L., Petersen, J. E., Hoogeveen, Y., Bamps, C., Burfield, I., van Swaay, C., (2008). *High nature value farmland in Europe. An estimate of the distribution patterns on the basis of land cover and biodiversity data.* JRC Scientific and Technical Reports EUR 23480 EN. Luxembourg: Office for Official Publications of the European Union.

Rudnick, D. A., Ryan, S. J., Beier, P., Cushman, S. A., Dieffenbach, F., Epps, C. W., et al. (2012). The role of landscape connectivity in planning and implementing conservation and restoration priorities. *Issues in Ecology, 16,* 1–23.

Sharp, R., Tallis, H. T., Ricketts, T., Guerry, A. D., Wood, S. A., Chaplin-Kramer, R., Nelson, E., Ennaanay, D., Wolny, S., Olwero, N., Vigerstol, K., Pennington, D., Mendoza, G., Aukema, J., Foster, J., Forrest, J., Cameron, D., Arkema, K., Lonsdorf, E., Kennedy, C., Verutes, G., Kim, C. K., Guannel, G., Papenfus, M., Toft, J., Marsik, M., Bernhardt, J., Griffin, R., Glowinski, K., Chaumont, N., Perelman, A., Lacayo, M. Mandle, L., Hamel, P., Vogl, A. L., Rogers, L., Bierbower, W., Denu, D., & Douglass, J. (2016). *InVEST +VERSION + User's Guide. The Natural Capital Project*, Stanford University, University of Minnesota, The Nature Conservancy, and World Wildlife Fund.

Smith, A. C., Harrison, P. A., Pérez-Soba, M., Archaux, F., Blicharska, M., Egoh, B. N., et al. (2017). How natural capital delivers ecosystem services: A typology derived from a systematic review. *Ecosystem Services, 26,* 111–126. https://doi.org/10.1016/j.ecoser.2017.06.006.

Snäll, T., Lehtomäki, J., Arponen, A., Elith, J., & Moilanen, A. (2016). Green infrastructure design based on spatial conservation prioritization and modeling of biodiversity features and ecosystem services. *Environmental Management, 57*(2), 251–256.

Star, S. R., & Griesemer, J. R. (1989). Institutional ecology, 'Translations' and boundary objects: Amateurs and professionals in Berkeley's Museum of Vertebrate Zoology. *Social Studies of Science, 19,* 3387–3470.

Tscharntke, T., Batáry, P., & Dormann, C. F. (2011). Set-aside management: How do succession, sowing patterns and landscape context affect biodiversity? *Agriculture, Ecosystems and Environment, 143*(1), 37–44.

Zulian, G., Polce, C., & Maes, J. (2014). ESTIMAP: a GIS-based model to map ecosystem services in the European Union. *Annali di Bottanica, 4,* 1–7. https://doi.org/10.4462/annbotrm-11807.

Zulian, G., Maes, J., & Paracchini, M. L. (2013). Linking land cover data and crop yields for mapping and assessment of pollination services in Europe. *Land, 2,* 472–492. https://doi.org/10.3390/land2030472.

Carlo Rega holds a M.Sc. in Environmental Engineering (Politecnico di Milano, 2002) and a Ph.D. in Spatial Planning and Local Development (Politecnico di Torino, 2008). He has been working as post-doc fellow at Politecnico di Torino from 2008 to 2013. He is currently working as Scientific Officer at the Directorate for Sustainable Resources of the European Commission Joint Research Centre (Ispra, Italy). His research interests include spatial and land use planning, mapping and assessment of ecosystem service, Geographic Information Systems, environmental assessment and agricultural policy. He has authored several publications in books and peer reviewed journals on these topics.

Chapter 7
Spatial Dimension of the CAP: Territorial Congruence, Supposed and Explicit Territoriality

Enrico Gottero and Claudia Cassatella

7.1 CAP and Planning: Evidence of Exchanges, Interactions and Alliance

The profound recent change in the agricultural and rural world requires careful and rigorous reflection on what is happening in our countryside and cities, especially in terms of ecosystem services. The role of agro-ecosystems in optimising the flow of supplies, support and regulation services, as well as the consequent production of economic, social and cultural benefits for mankind, has long been widely debated and acknowledged in the scientific, political and institutional context, both nationally and internationally. The approach refers to the tangible concept of *landscape services* (Willemen et al. 2008; Termorshuizen et al. 2009) is, instead, less common and records less interest than material flows of water, food and energy. However, the agrarian landscape could perform several functions and contribute to the preservation of soil, maintenance, management and control of forests, transmission of a place's identity, and preservation of cultural (material and immaterial) diversity. It is also of particular importance to safeguard aesthetic value, increase recreational and tourist use, and for effects on the economic system (tourism, typical products market, etc.). It also offers cities an opportunity to take

Although the manuscript is the result of the joint work of both authors, the Sects. 7.2.1, 7.2.3 and 7.3 are mainly attributed to the work of Enrico Gottero. These sections were also presented by this author at the XXXVII Italian conference of regional science (AISRe) (Cagliari, 2017, 19–21th September).

E. Gottero (✉)
IRES Piemonte, 10100 Torino, Italy
e-mail: e.gottero@ires.piemonte.it

C. Cassatella
DIST, Interuniversity Department of Regional and Urban Studies
and Planning, Politecnico and Università Torino, 10125 Turin, Italy

E. Gottero (ed.), *Agrourbanism*, GeoJournal Library 124,
https://doi.org/10.1007/978-3-319-95576-6_7

advantage of services and agricultural products designed to meet the needs of the urban market (direct sale of local products, farm holidays, etc.).

In this context, the role of *Common Agricultural Policy* (CAP) has gradually become very important. Although originally conceived for other reasons (support for farmers' income and market regulation), the CAP has progressively expanded its scope and gradually strengthened the territoriality of its tools. Growing interest in the environmental dimension has given rise to a "driving" CAP in terms of transformations that involve the landscape. The original CAP framework, mainly studied to enhance productivity, has made room for a strongly environmental matrix. The succession of many reforms, including *Mac Sharry* (1992), *Fischler* (2003) and *Health Check* (2008), has gradually led the CAP to integrate European environmental objectives and, consequently, to concretise the European economic intervention for the rural landscape. Today, the CAP is a tool that can support and improve rural landscapes. It gives farmers an active role in managing environmental and cultural resources.

Currently, CAP 2014–2020 comprises two pillars: (1) *market policies and direct payments*—in order to support the agricultural sector and productivity—and (2) *rural development* that aims to strengthen rural areas and preserve biological and cultural diversity. However, the territorial approach of the second pillar, introduced by *Agenda 2000* (CEC 1997), has progressively brought the CAP closer to the rural landscape, on the basis of some fundamental principles. The first concerns "ecosystemic conditionality", and establishes that compliance with environmental protection standards (Birds and Habitats Directives) and food quality standards, besides low discharge of chemical waste are crucial requirements to obtain public support. Another important aspect concerns the concept of "diversification". Depopulation and abandonment of the countryside have forced the CAP to support income sources other than conventional farming. Lastly, based on the concept of *multifunctionality*, in order to survive, agriculture cannot solely devote itself to supplying raw materials. It will also have to provide services, such as landscape conservation and enhancement, farmland maintenance, transmission of the identity of places and maintenance of biocultural diversity, as well as recreational use of rural space (tourism, hospitality, etc.).

Although restrictive environmental constraints and ecosystemic conditionality also apply to direct payments, which are the most important in terms of allocated resources, the first pillar of the CAP has only recently mentioned an indirect link with the landscape, mainly favouring production. However, the new CAP 2014–2020 regulations have introduced significant changes to the mechanism that underpins the first pillar, notably through the concept of "greening". Direct payment is, therefore, subject to compliance with some *climate-friendly* and *environmentally-friendly* agricultural practices that are far more restrictive than previous ones, such as *crop diversification*, the *maintenance of existing permanent grasslands* and the inclusion of an *ecological focus area* (EFA).

Unlike the landscape contribution of the first pillar of the CAP, which is still not "visible", rural development policies seem to present the largest number of relationships with the landscape and the territory. Common European strategies in this sector were implemented through *Rural Development Programmes* (RDPs) based on a *Common Strategic Framework* (CSF) and three objectives (competitiveness, sustainable management and territorial development). RDPs are divided into six priority areas or *focus areas*, which contain the so-called measures that are implemented both through regional competitions and also through various forms of networking (rural networks, PEI, local development, LAG, etc.). These measures—such as agri-environment-climate payments (Measure 10)—mainly have positive indirect effects on the rural landscape, especially concerning the ecological, perceptual-visual and land use dimensions. However, such tools must be used with the utmost care and awareness, within a larger integrated and *site-specific* landscape project that can qualitatively orientate the multiplicity of measures promoted by RDPs.

Thus, there is a "toolbox" to govern and design the rural landscape, but it is heterogeneous and extremely complex. It includes several tools that have not been encoded as yet, and which mainly focus on the principle of integration between landscape and rural development, as well as on the mechanism of incentives. Landscape policies can, therefore, be implemented not only through RDP measures but also through the rules of *conditionality*, *greening*, *equivalent agricultural practices*, *Strategic Environmental Assessment* (SEA) and many *place-based* implementing devices that stem from *agreements* or *territorial agro-environmental schemes*, *Integrated Supply Chain Projects* (PIF), *Local Development Plans* (LDP), pilot projects, etc. We can consider, for instance, the many local agri-environmental agreements or schemes, such as those contemplated by the Marches Regional Administration's 2007–2013 Rural Development Programme and France's *Mesures agro-environnementaux territorialisès,* or the many place-based initiatives proposed by *Local Action Groups* (LAGs), river contracts, *Ireland's quiet areas, Sustainable Agriculture Contract*s or agri-urban programmes in France, the Dutch *Eco-Coops*, and *Environmental Stewardship Schemes in the United Kingdom* (Fanfani 2013; Jeanneaux et al. 2011; Rega 2014; Rovai et al. 2008; Scolozzi et al. 2012). These are mostly participatory tools and trades that leverage on heritage and local knowledge, considered as an effective reading and project key to re-establish the rural landscape's value.

Hence, we cannot operate on the landscape without dealing with agriculture. Although such a link is indispensable, it would be less trite to theorise an alliance with territorial and landscape policies as a regulatory background, where agriculture is the main tool supported by financial devices to preserve and enhance the landscape. However, this is an experimental hypothesis that has to deal with many obstacles and challenges (Cassatella and Gottero 2016).

In Italy, for instance, current sectoral separatism is one of the most important issues. On the one hand we find the CAP and, in particular, RDPs and, on the other, we notice regional territorial/landscape plans (RTPs, RLPs, etc.). Despite partially pursuing the same objectives, these tools have communicated with obvious difficulties, through different "languages". This is also due to the lack of functional integration between plans and programmes. Spatial plans are limited to regulating settlement phenomena, where possible, by assigning strategic goals without the financial tools required to implement them (Cassatella 2015; Gottero 2016a). Conversely, agricultural policies, however strong in economic terms, often either ignore or reject the strategies and tools of territorial plans (cognitive framework, monitoring systems, etc.). This stems from a lack of knowledge of the landscape context in which agricultural policies operate, and also from a generalised lack of awareness of territorial and landscape implications. Unfortunately the outcome features inconsistent interventions, poor efficiency in public spending (resources are expended in areas that either do not need them or which are not a priority) and, at times, loss of not replicable resources, identity and cultural-historical heritage.

Various institutions, researchers and scholars from heterogeneous disciplinary fields have long attempted to follow the integration pathway. Some of the most recent studies (ECA 2011; Cassatella and Gottero 2016; Gottero 2016a, b; Gottero and Cassatella 2017; Piorr et al. 2009; Prager et al. 2012; Rega 2014) focus on the central role of agricultural policies in implementing strategies drawn up by urban planning schemes, and in addressing functional gaps through a common and coherent language, hypothesising a strongly *site-specific* CAP.

In this chapter the authors will make use of some of the most recent studies conducted in Piedmont (Italy) to describe methodological issues of CAP spatialisation, as well as the results of an initial assessment of the territorial and landscape contribution of both pillars. Finally, in the last section, after briefly discussing the limits and potential of the method and the results recorded, the authors will make their conclusive reflections on open issues and other possible developments or research outlines.

7.2 Spatial Evidence of the CAP: Materials, Methods and Results from the Piedmont Case Study (Italy)

In the following pages, the authors will discuss some of the results of the Piedmont case study, taking into account the complexity and relevance of both CAP pillars. The rural landscape of Piedmont, a north-western Italian region that stretches from the Alps to the plains of the Po valley, is highly agricultural. Farmland covers more than half of its surface, and large areas are given over to woodland and pastures. Hence, it is extremely difficult to single out features that can be regarded as typical. It could be defined as a multitude of landscapes where stretches of traditional countryside coexist with areas that are intensely cultivated with arable crops and vineyards (Cevasco 2013). This is also inferred by the renewed *Regional Landscape Plan* (RLP), which was introduced in 2009 and recently updated (2015). It is also one of the most important sources not only to identify specific territorial features but also to define the components of rural landscape. The reference to a regional source is not unusual. Data on spatial and landscape elements (and their tendencies) are usually collected at this scale. In addition to these, land use and coverage (Corine Land Cover, Urban Atlas, Regional Forest Map, etc.), as well as alphanumeric and spatial data related to the RDP and contained in regional datasets (operational DB of RDP, data warehouse on rural development, etc.), are important for the subsequent stages of information processing.

For these reasons, the first part will provide a preliminary analysis (ex-ante) of some data of the first pillar, especially in view of an ongoing assessment of its spatial and landscape effectiveness. The second section will, instead, enlarge on the spatial congruence of the *Piedmont RDP 2007–2013* regarding the landscape objectives of RLP (ex-post), while the last section analyses territorial priorities assigned to some RDP operations. It reflects on the supposed territorialisation of *Piedmont RDP 2014–2020* and RDP efficiency, especially in terms of environment and landscape in the ex-ante stage.

7.2.1 Spatial-Explicit of CAP 2014–2020 in the Piedmont Region

To date, the spatial dimension of the CAP is almost exclusively an RDP domain. However, the financial importance of the first pillar has progressively revived interest in spatial implications, also in light of the latest innovations introduced by the *European Structural and Investment Funds* (SIEs). Unfortunately, research in this direction remains modest and often evanescent, also for the lack of information.

Direct payments are a substantial part of the total CAP budget, approximately 70%. Although increasingly subordinated to restrictive environmental constraints and to compliance with ecosystemic conditionality (reduction of chemical discharge, European habitat and bird directives, food quality, etc.), they are still anchored to the productive sphere. However, recent EU regulations[1] have rapidly reinforced the "green" component of this supportive system. In addition to basic payments and retribution both to young farmers and to areas presenting natural constraints, the CAP 14–20 has added the "greening" component based on a set of significant additional environmental requirements (Vanni 2013), such as *crop diversification*, *maintenance of permanent grassland* and the *creation of EFA*. The latter, in particular, concerns *set-aside* practices, areas presenting traditional landscape features, such as hedges, tree lines, stone walls, terraces, as well as afforestation areas, intercalary crops, etc.

In this regard, the Piedmont case study highlights some core issues (Table 7.1). The data analysis reveals that greening has involved 38% of the total area

[1]See: Regulation (EU) No 1307/2013 of the European Parliament and of the Council of 17 December 2013, establishing rules for direct payments to farmers under support schemes within the framework of the Common Agricultural Policy and repealing Council Regulation (EC) No 637/2008 and Council Regulation (EC) No 73/2009. http://eur-lex.europa.eu/legal-content/EN/TXT/PDF/?uri=CELEX:32013R1307&from=it (Accessed 06/02/2018).

Table 7.1 Areas associated with direct payments

Payment type	Total area associated (ha)
Diversification	479,211.23
Ecological focus area (EFA)	45,796.53
Permanent meadows	165,001.19
Total area associated with greening payment	690,008.95
Total area associated with direct payments	1,118,654.95

Source Author's own elaboration on CSI Piemonte alphanumeric data, 2016

associated[2] with direct payment (€ 190 MLN for 2016), in other words 70% of the total budget allocated to *Measure 10 of the RDP for the period 2014–2020.*[3] From a geographical point of view, the area associated with greening payment seems to mainly involve arable land between Turin and Cuneo, and near Alessandria, Novara and Vercelli, which feature intensive farming. The southwest quadrant of the region is also covered by greening payments (probably pastures), while the hills between Langhe and Monferrato, as well as the low Valsesia are less significant (Fig. 7.1). The arable crop area, where diversification is applied with two or three different substitutive crops, is the largest part of the surface area associated with greening (70%), which also includes part of Piedmont's *Used Agricultural Area* (UAA). The area maintained for meadows and grazing is, instead, just under 10% of the total area associated with direct payments, and is less significant than the previous one. The EFA analysis, instead, revealed that this dimension of greening has primarily involved nitrogen-fixing crops and fallow lands, whereas practices designed to re-establish traditional (linear and punctual) components of the rural landscape are less significant, compared to the previous ones. In this regard, although only 5% of the arable land area (more than 15 ha) is mandatory for this purpose, about 800 km of linear elements, such as hedges and ditches, have been constructed and/or maintained through EFAs.

7.2.2 *Spatial Congruence of the Piedmont RDP 2007–2013*

In this case the study adopts an approach that considers the rural landscape an organised set of interacting aspects. Our theory of implementing constraints and rules, and also of practically defining strategies laid down by regional plans, relies on a coordinated spatial initiative rooted in the CAP's incentives mechanism.

[2]This is the surface on which the payment is calculated. It does not correspond to the real area invested by each intervention or agricultural practice. For these reasons sometimes its value could seems overestimated.

[3]The financial plan of the Piedmont RDP 2014–2020, approved in March 2017, assigned about 263 MLN to this measure.

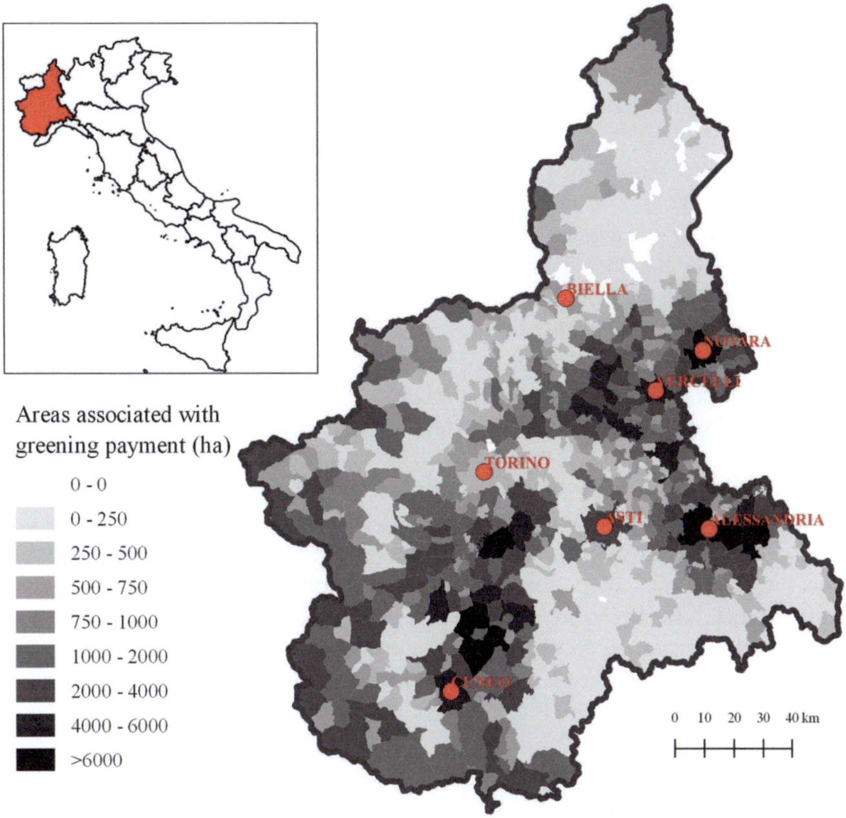

Fig. 7.1 The geographical distribution of the *greening or ecological payment* of the first pillar of the CAP 2014–2020. *Source* Author's own elaboration on CSI Piemonte alphanumeric data, 2016

Accordingly, while maintaining the rural landscape as our unifying theme, the following paragraph will focus on interactions between the 2007–2013 *Rural Development Programme* and guidelines contained in documents that make up the 2009 *Regional Landscape Plan*. Our aim was to understand what landscapes had been supported by the 2007–2013 agricultural policies, and whether measures adopted by the RDP (Piedmont Regional Authority 2012) have been consistent with landscape priorities advanced by the RLP (Piedmont Regional Authority 2009). To this end, areas covered by funding were mapped by georeferencing data concerning the beneficiaries of RDP. We were thus able to compare various aspects of the region's rural landscape with the measures and actions fielded by the programme, whose results can be assessed ex-post (Spaziante et al. 2013) to determine their consistency and interaction between sectoral plans. We were also able to assess the efficacy of individual measures contemplated by the RDP. This analysis was intended to check for possible symmetries, examine the consistency between sectoral plans, and evaluate the efficiency of the region's public policies for the agricultural landscape. Based on GIS tools, the approach used overlay maps of areas supported by the RLP to highlight the intensity of CAP aid and the spatial distribution of the support. Through the analysis and representation of geographical data, this stage produced a *congruence analysis model* that can also be used for the 2014–2020 RDP.

The proposed model thus overlays the RLP and RDP. The former is a tool, whose primary goal is to protect and enhance the natural, cultural and landscape heritage. The RLP is Piedmont's regulatory framework for the countryside, and also provides useful guidelines for multisectoral policies that concern the region's entire territory and links with the European context. The plan divides Piedmont into 76 landscape zones and 535 sub-areas or landscape units on the basis of the recursive and unitary nature of the environmental and cultural patterns that emerged (Piedmont Regional Authority 2009). Hence, the structure and identity of these areas is rooted in certain common features (geomorphological, natural, historic, cultural, cultivational, etc.), which rank them as homogeneous units. The fact that the RLP is organised by zones suggests that the first stage of organising knowledge about the rural landscape should identify the *strategic objectives* and *values promoted* for each landscape unit, precisely the implementation provisions for rural areas presenting high biopermeability, high agronomic interest and specific landscape interest, besides *traditional features,* such as forges, mills, Alpine meadows, rural pastures, types of cultivation that convey a site's distinctive identity (rice fields, vineyards, etc.), hedges and row plantings, terracing, paths, mule tracks, country roads and so forth (Fig. 7.2). As for the RDP, the first step focused on

Fig. 7.2 The historical viticulture of the Monferrato Astigiano. *Photo* Enrico Gottero

identifying actions and initiatives that affected the landscape, and then on distinguishing the potential direct and indirect impact on the landscape of actions implemented by December 31, 2013 (Table 7.2). Lastly, data from the CSI Piedmont regional rural development monitoring data warehouse were georeferenced and overlaid on the landscape zones indicated in the RLP, making it possible to check spatial targeting of the 2007–2013 RDP against the objectives and values promoted by the RLP. The process yielded a matrix that can be consulted in two ways, by "measure" and by "landscape zone", and allowed a dual evaluation in terms of "intensity" and "congruence". This breakdown allows to identify landscape zones that are most affected by the RDP's "landscape actions" (spatial coverage), as well as the average intensity of initiatives in the region's entire territory. At the same time, it ascertains the RDP actions' degree of spatial congruence with the strategic framework of the RLP, and the overall congruence of the RDP's landscape initiatives with goals and values contemplated by the RLP for each landscape zone. Thus, the measures' *spatial coverage* is evaluated by assigning a score from 0 to 3 (none—high) to each measure's level of concentration. *Congruence*, on the other hand, is evaluated by assigning a score from 0 to 2 (none, incorrect, correct) to each measure's spatial correspondence with landscape priorities or emergencies indicated in the RLP. In addition, the matrix devotes particular attention to the intensity of the action, as well as to the reciprocities with the system of rural areas of high agronomic interest (Article 20 of the RLP), rural areas of specific landscape interest (Article 32 the RLP), and to the score assigned in relation to the traditional features identified by the matrix. The model also refers to specific criteria associated with each individual initiative (strategic objectives and values promoted by the RLP), where the intensity of the action is deemed positive for initiatives having a positive effect, a prerequisite to guarantee uniformity, and is an unfavourable condition for actions that have a negative or unclear impact. This approach takes into account not only the degree of concentration of RDP actions, but also the failure to involve specific rural areas of landscape interest identified by the RLP. This negative factor jeopardises the system of relationships between spatial priorities of both regional policy documents.

The 2007–2013 RDP for Piedmont chiefly promotes indirect actions with potentially positive effects, mostly involving the visual and land use aspects of the landscape, in particular through Axis II (*Improving the environment and countryside*) and agri-environmental payments. For the major part, these actions enhance and improve landscape quality, while measures that can potentially put pressure on the landscape basically concern the construction of buildings and installations for the production of renewable energy (photovoltaic, biogas, etc.), especially those entailing variations in land use and changes in the visual appearance of certain points or entire areas (*Axis I—Improving competitiveness of agriculture and forestry sectors*). By contrast, other measures, particularly those entailing the afforestation of farmland, are apparently positive but can conceal negative effects in terms of losing features that convey the traditional rural landscape's identity. In other words, though an increase in wooded areas might bring about a real improvement in the quality of the landscape, it may at times cause the loss of the

Table 7.2 Piemont 2007–2013 RDP actions and initiatives affecting the rural landscape (Authors elaboration)

Measure	Actions	Potential impacts
121	Construction of new farm buildings	Generalized visual impacts, areal visual impacts, farmland consumption
121.1.1–HC	Installations for production of renewable energy	
121–121.1.2A	Anti-hail fixed installations	
121.2.1.A–HC	Water storage	
121.E	Construction of driers	Specific visual impacts
121.1.2.B–HC	Works to prevent landslide/erosion	Generalized visual impacts, impacts on hydrogeological control
214.3	Increase organic carbon of soil	Indirect impacts on hydrogeological control
125.1–125.3.1	Improvement of forestry and rural roads	Generalized visual impacts, linear impact on fruitive dimension
227	Forestry investments for tourist-recreational purposes	Specific, linear and areal visual impacts, linear impact on recreational dimension, impact on historical and cultural dimension
313.1	Construction of footpath network	
313.2B	Construction of recreational infrastructures	
132	Improving food quality systems	Impact on historical and cultural dimension
211/214.6.2	Support for agricultural activities (only mountain areas)	Direct impacts of institutional action
214.6.1	Support for pastures	Areal visual impacts, impact on historical and cultural dimension
214.1	Integrated production	Areal visual impacts, impact on historical and cultural dimension
214.2	Biological production	
214.4	Conversion of arable land to permanent grassland	
214.7	Elements of agroecosystem	
214.9	Interventions for biodiversity of paddies	
H-REG.2080	Afforestation of agricultural lands	Areal visual impacts, impact on areas affected to erosion or hydrogeological problems, impact on historical and cultural dimension
221		
311	Remodeling of rural buildings	Specific, linear and areal impacts on recreational dimension, impact on historical and cultural dimension
323	Conservation and regeneration of rural heritage	

area's traditional character. Likewise, while woodland may increase ecological diversity and visual variety, it cannot be taken for granted that afforestation is necessarily a means of protecting a place's historical and cultural identity (Agnoletti 2013). A significant contribution towards enhancing the historical-cultural heritage is also provided by actions that increase the use value of the countryside by promoting productive and recreational activities, as well as scenery preservation (*Axis III—Rural quality of life and diversification of rural economy*) (Gottero 2016a).

A number of other considerations stem from the evaluation matrix, particularly as regards the measures' spatial target. The *intensity* and spatial coverage of "landscape actions", i.e., only actions that can have potential effects (whether positive, negative or indeterminable) on the landscape can be analysed on the basis of measures. They reveal that initiatives, which have a negative effect on the rural landscape—most of which are Axis 1 investment measures—generally have low intensity, or a weak impact on the region's countryside. In this case, the construction of new farm buildings and installations for the production of renewable energy are the main sources of pressure on the rural landscape. As for actions with positive effects on the landscape, the analysis reveals a system that is chiefly medium-density, whose highlights are agri-environmental measures and, in particular, subsidies for marginal mountain areas, as well as organic and integrated farming. A less significant contribution is made by measures falling under Axis III of the RDP, i.e., initiatives to recover rural architectural heritage, infrastructuralisation of the network of paths and recreational trails, and tangible and intangible efforts to protect and restore the rural heritage, all of which are low intensity. The first question thus regards the concentration of "landscape actions" which, as we see in Fig. 7.3, affect more than one geographical group in the entire region. In this sense, the RDP's contribution to landscape is particularly strong in the widely known UNESCO World Heritage-listed wine growing areas in the *Basse Langhe* and *Monferrato Astigiano*. These areas are where the RDP most likely expresses the greatest interest in landscape, while *Alpine areas of northern Piedmont* appear to have been ignored, despite their considerable value. In general, intensive assistance is particularly provided in the plains and hills of south-east Piedmont.

In this sense, the *congruence analysis*, conducted first on the basis of measures, found a system of landscape actions that is moderately consistent with the RLP, especially for agri-environmental measures. It is also clear that there is a singular symmetry between certain actions, where high intensity corresponds to limited consistency (e.g., Measure 121 and measures for mountain areas). Once again, Axis III measures are the least consistent with the aims of the RLP (measures 311 and 323). Lastly, the matrix allows to identify areas where the RDP's spatial targeting is effective or ineffective in terms of preserving and enhancing the countryside. In this sense, the geographical distribution of "landscape actions" appears to be less uniform than the allocation of funding. Zones where the RDP's "landscape actions" are most consistent with the system of goals and values promoted by the RLP are situated in the mountains, although they generally have a low intensity. There is also a high degree of consistency in the hill country, while interconnections of zones to the southwest of the regional capital, Turin, are much

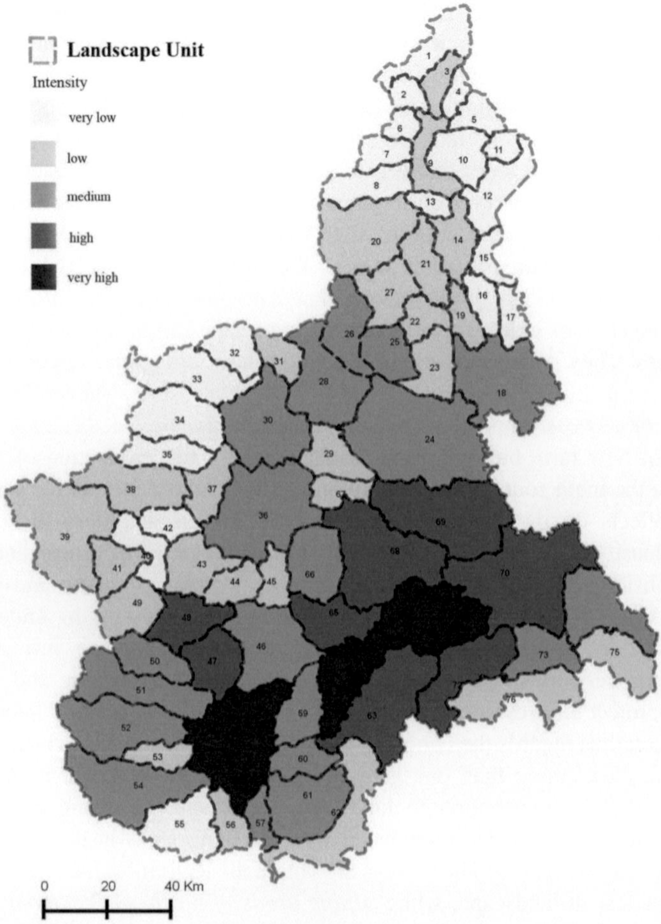

Fig. 7.3 Intensity of RDP landscape actions (Authors elaboration)

weaker. The evaluation matrix indicates that the intensity of aid does not always correspond to correct spatial targeting (Fig. 7.4). This underscores the importance of converging spatial objectives which, if poorly formulated, can limit the effectiveness of regional public policies "for the rural landscape" (Ibidem).

7.2.3 Supposed Territorialisation of Piedmont RDP 2014–2020

The analysis of spatial implications of RDP 2014–2020 of Piedmont (Piedmont Regional Authority 2015a) has highlighted the vocation and gaps of a tool which,

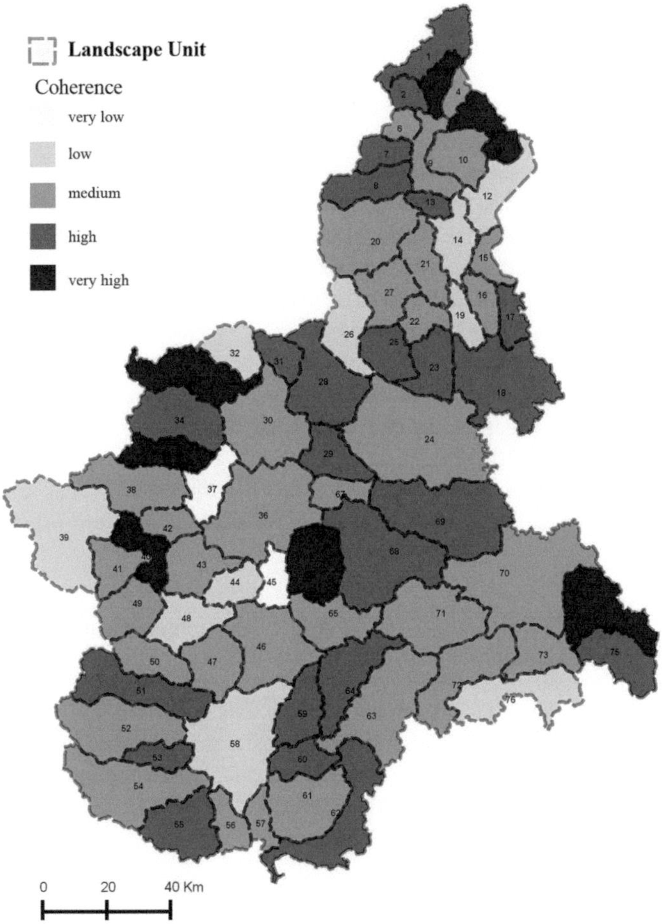

Fig. 7.4 Congruency of RDP landscape actions (Authors elaboration)

however, seems to be more equipped to sustain territorial reasoning and indications than the first pillar. In order to check the supposed geography of the RDP[4] and show the RDP's territorial potential (negative and positive), it has systematised priorities, criteria and any specific territorial indications contained in the RDP, and also classified spatial and qualitative eligibility conditions. Most RDP operations, which correspond to specific financial allocations (see the RDP section on financial plan), show one or more territorial priorities in the ex-ante phase (and then in their implementation through public competitions and selection and/or eligibility criteria)

[4]It is not an effective ex-post spatial distribution of RDP measures or of territorial priorities defined by implementation phase (public competitions). These priorities are solely defined by RDP in the ex-ante stage.

that, in most cases, partially overlap. However, sometimes territorial criteria are uncertain, such as, for instance, the *Regional Ecological Network* (RER) and *High Natural Value areas* (HNV), which are still being defined. Others can neither be clearly identified nor located (areas with high chemical discharge), while some of them have required specific research, selection and systematisation of numeric and spatial data (such as surfaces cultivated through biological methods). In other cases, the geographic information used had obvious time constraints, such as arable land and low carbon organic areas. However, the explicit territorial indications of the RDP 2014–2020 of Piedmont were subsequently reported with the operations and financial allocations that determined the average value of resources within each spatial priority (Table 7.3). These values were subsequently associated with geographic information through GIS tools, by overlapping each layer, including *Regional Socio-Economic Homogeneous Areas* (AITs) and *Landscape Areas* (APs), and by using a generic spatial grid (5X5). Finally, to verify possible symmetries or other relationships, spatial analysis was further supported by some territorial and landscape "reading matrices" defined by the RLP, namely *Rural Areas of Landscape Interest* (Article 32) and *Areas of High Agronomic Interest* (Article

Table 7.3 The explicit territorial indications of the RDP 2014–2020 of Piedmont with the average value of financial allocations for each spatial priority (Author's own elaboration)

Area	Average value of financial allocations (%)
Urban and periurban areas	0.27
Areas of intensive agriculture	0.24
Intermediate Rural areas	4.12
Intermediate rural areas with natural constraints	6.46
Rural areas with development issues	15.7
Protected areas	4.82
Natura 2000	10.2
Fluvial bands	0.47
hydrogeological risk areas	1.47
Forest	0.21
Ecological Network Areas	3.2
Safeguard areas from water uptake	1.9
Nitrate vulnerable zones	4.46
Pesticide vulnerable zones	3.04
Hill	0.24
Mountain	5.98
Lowland	1.98
Arable crop	1.62
Area with low organic matter in soil	0.69
Areas for organic production	1.51
Local action group (LAG)	6.27

20) (Piedmont Regional Authority 2015b). The former are characterised by particular natural and/or cultural components, including heterogeneous rural and agroforestry systems, traditional arable land and settlements, while the latter basically include high-agronomic capacity areas, as well as traditional or certified production.

Spatial analyses carried out on supposed territorialisation of Piedmont RDP 2014–2020 have shown some common issues relative to the lack of "reasoned territorialisation" of measures. The first one concerns the number of interventions with an explicit territorial reference (only 38 out of 64). This is a significant amount, considering the overall budget allocated to the RDP. In addition, it should be considered that not all operations can have a spatial reference, such as measures concerning professional training, advisory services, young farmers, etc. Moreover, approximately 75% of the budget is territorialised (about € 790 MLN), which means that the RDP is not completely devoid of any specific indications of this kind. Finally, it must be said that these elements neither assess the validity of territorialisation nor do they represent the actual territorialisation of RDP actions.

The supposed territorialisation of Piedmont RDP 2014–2020, carried out by overlapping different layers and spatial information, highlighted certain spatial factors. The Fig. 7.5 shows the prevalence of support for mountain areas. This underscores a first territorial issue, since agriculture in that area is considered marginal and presents several development problems. Hence, the RDP is mainly a support mechanism in less-favoured areas defined by altimetric criteria. In this sense, hills and plains are still left out although, especially from an environmental point of view, they have to face issues that, at times, affect the development of the whole region (e.g., intensive farming, livestock, etc.). Some vital areas, not only in terms of production but also of development of increasingly important complementary activities, such as rural tourism and local products (i.e. the plains and landscape unit of Alessandria, Cuneo, Novara and Vercelli), seem to be poorly supported by the RDP. Finally, urban centres are left out of this rationale, confirming the trend of recent years which, without regard to the progressively acquired importance of urban and peri-urban agriculture (Gottero 2016b; Lohrberg et al. 2016), has mainly financed rural areas.

In addition, spatial analyses based on grids and other layers have underscored some issues that contrast with previous ones, especially in terms of spatial relations with the RLP. The Fig. 7.6 shows that the covered cells of agronomic interest are mostly located in the plain between *Turin and Cuneo*, while rural areas of landscape importance covered by the RDP are largely located in the southwestern and northern grids. Significant areas for both purposes are the *Turin Metropolitan Area*, as well as the *Novara* and *Vercelli* plains. Finally, the Fig. 7.6 shows that culturally important areas, such as the UNESCO hills between *Langa and Monferrato*, are often not covered by the RDP. Hence, areas targeted by measures presenting explicit territorialisation in excess of 1% aid, cover only a minor part of regional rural areas of landscape interest, which would undoubtedly deserve more attention.

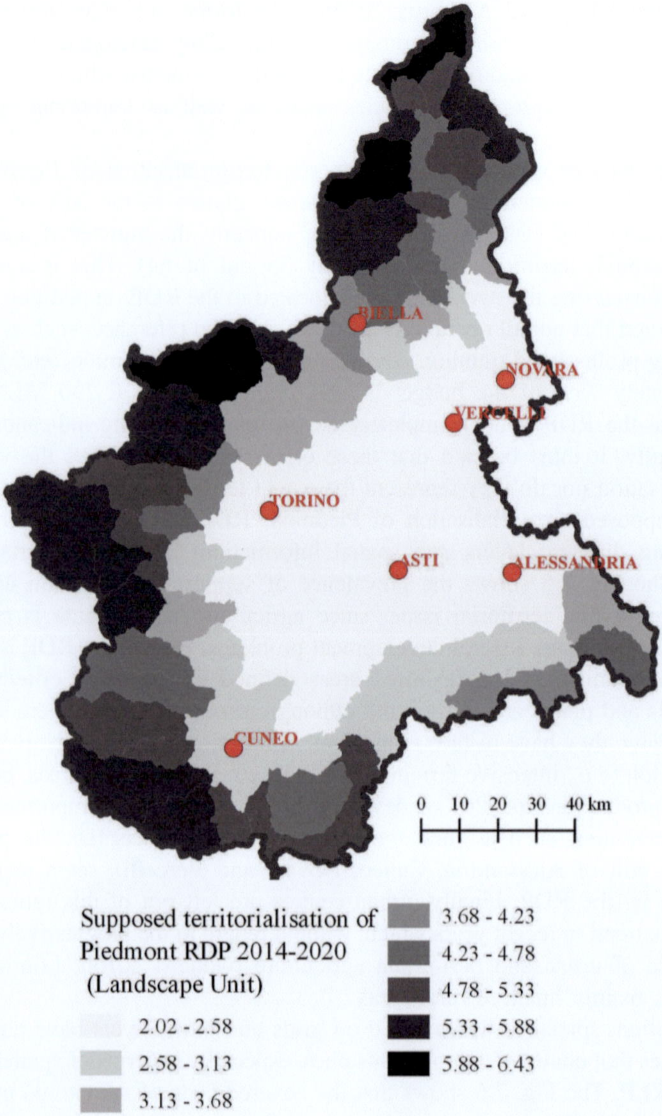

Fig. 7.5 The Supposed territorialisation of Piedmont RDP 2014–2020 for Landscape units (Author's own elaboration)

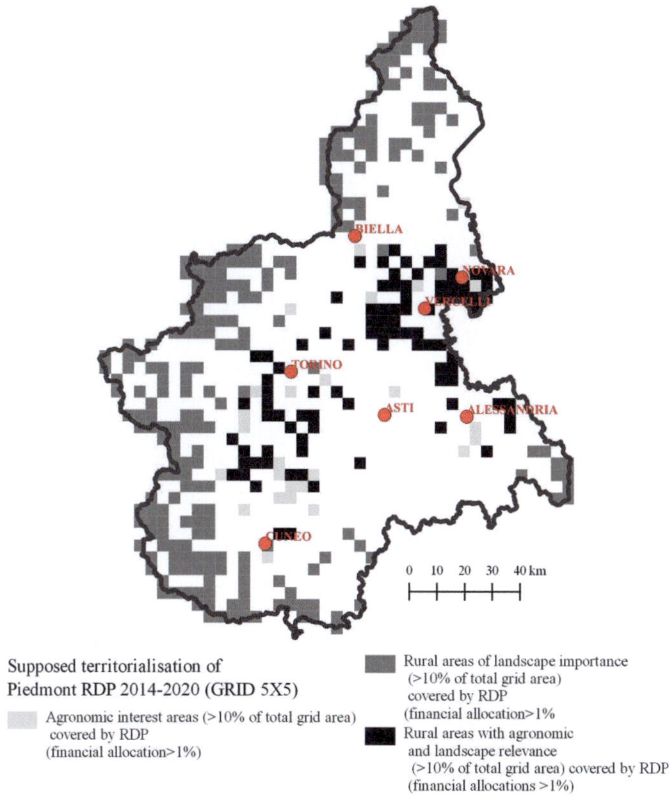

Fig. 7.6 The Supposed territorialisation of Piedmont RDP 2014–2020 for grid 5X5 (Author's own elaboration)

7.3 Discussion

The results of the analysis carried out on the Piedmont case study also underscored some of the territorial and landscape features of the CAP and, in particular, of the RDP. In this respect, the outcome of this case is particularly important for certain issues:

– *applicability and validity of methods:* despite not being exhaustive, the various illustrated methods could be used in different stages of both CAP pillars, from programming to reporting. These tools guide decisions, verify consistency and interaction with other sectoral plans (ex-ante, on-going and ex-post), evaluate effects, ensure correct and fair distribution and rationalisation of financial resources (also in terms of performance of RDPs), and increase awareness of the CAP in terms of ecosystemic value system and services associated with the rural landscape. These methods could be replicated in several European and regional contexts, although they need a structured and verifiable information system, and cannot be extended to any operations or practices supported by the CAP;

- *effectiveness and efficiency of the CAP:* achieving landscape and territorial objectives, and ensuring efficient public spending are some of the critical points that surface from this study, not only for Piedmont, but for the entire domain of European agricultural policies. The hypothesis of providing concrete solutions for strategies that have already been defined at the regional level by urban and regional planning through RDPs—the theory is being tested in many European and regional contexts (see also: Cassatella and Gottero 2016)—while also enriching cognitive frameworks and performing contextual analyses of RDPs through information from RLPs, seems to be essential for correctly functioning development policies. At the same time, the rationale which underpins extensive distribution of considerable public resources seems to be disconnected from territorial criteria. Failing territorial awareness, *site-specific* needs and knowledge of a site's value system, implementation of the CAP will probably only be partially efficient. Finally, we need to study the effectiveness of some "environmental" actions defined by RDPs. Although data on the first pillar are currently undergoing validation, the analysis carried out on both pillars showed that the covered areas of green payment are often comparable to agri-environmental measures. This means that implementation of the CAP must be monitored, verifying the effects and, perhaps, the "levers" of the first pillar, also rationalising some of the resources allocated to the RDP;
- *spatial priorities of RDPs:* especially the ex-post evaluation shows the congruence of two instruments that were developed, although not formally, paying little attention to their mutual interaction. The assessment shows that this lack of coordination, which is particularly apparent in the RDP's spatial targeting, results in significant public resources being spent in a way that is both ineffective and inconsistent with the landscape and territorial objectives set by the spatial plan. The approach described offers a few certainties and many doubts, mostly arising from the lack of an actual link between the CAP and landscape planning, as we can notice by the fact that there are no specific, explicit landscape objectives, especially in the RDP. If, on the one hand, the landscape issue is relatively new in the history of common agricultural policies, so much so that it is often confused with the environmental dimension, it is now equally obvious that the CAP is of vital importance, particularly for the next planning cycle, as it makes the objectives outlined by the landscape plans tangible, especially in the case in question. The contribution of this tool is so great in terms of repercussions on the countryside that landscape aims and actions cannot be grouped with the generic environmental dimension. First and foremost, we need "formal recognition of value", expressed through targeted objectives and actions. It should reach beyond the current one-dimensional principle, which restricts the landscape to a mere, and often neglected, component of the environment, in favour of a broader holistic vision. However necessary it might be, this alone is not sufficient. Systematic attention must also be devoted to the current "toolbox" that is available to decision-makers, so that they can act on the principle of

intersectoral sharing of objectives. It is a matter of progress centred on strategic integration, on the common objective of "protecting and enhancing the countryside" and on seeking to bridge the regulatory gap or, in other words, the lack of landscape requirements to access first pillar funding and rural development support, which are consistent with the objectives of the RLP. It also means establishing new geographic allocation criteria for public spending based on spatial targets shared with the RLP. This is the criteria's trend, namely the consistency requirements a measure must meet to be considered as landscape-oriented, established by spatial targeting. Lastly, we must identify areas where there could be conflicts (or which are ineligible), and specify territorial priorities that take landscape emergencies and the objectives of the RLP into account, for instance through a recognition process shared by the areas involved. These aspects, together with formalisation of the RDP's landscape aims, are additional essential steps towards improving the effectiveness of regional public policies for the landscape. Finally, as in other Italian regions, territorial distribution of Piedmont RDP's should, therefore, reconsider the biological, cultural and landscape richness of areas that are often "scarcely marginal" but "sufficiently neglected" to affect socio-economic development. The classification based on the clear dichotomy between urban and rural seems to have been overcome. Consider, for example, the importance of these aspects, taking into account the eligibility issue of urban areas, as well as the criticality of the classification method itself. A *place-based* approach and *site-specific* priorities could reverse this trend and reaffirm the validity of other criteria that are not only bound to peripherality and marginality, but also to the uniqueness of certain landscapes that have been ignored to date.

7.4 Conclusions

The methods illustrated herein are intended to provide guidance for initiatives and to identify the territorial and landscape priorities of current and future agricultural policies. However, pilot projects are needed to validate this approach in order to concretely demonstrate what can be done and what the outcome will be. How can we capitalise on the input provided by agriculture, environment and landscape? How can we work together, reconciling the different goals and defusing potential conflicts, aided by indicator system and scientific monitoring, which accompany the path? How can we involve both the public and stakeholders? What can be obtained by accepting experimentation? Our proposed alliance between rural policies and landscape policies is a difficult prospect from a political standpoint, compared to the usual "scattershot" distribution of funding, and considering the jealousy shown by the different branches of local government. By establishing such an alliance, we will be able to contribute towards some of the objectives of landscape plans through funding earmarked for agriculture. However, RDP measures are on a voluntary basis. Why should the rural world participate? Because in Italy (and not only in Italy), the countryside's prospects for development are hinged, to an increasing extent, to the quality of the landscape, its attractiveness for tourism, its territorial branding capacity, and the economic value of products bearing its brand.

Hence, now we need to define eligibility criteria for funding applications with an eye to the areas' landscape features, and also focusing on specific areas, such as planting hedgerows and the like, where needed, to establish the regional ecological network, implementing afforestation to return degraded areas to health, maintaining or restoring clearings to ensure the continued legibility of the built environment, and creating paths and trails as part of setting up landscape networks. In these terms European rules are not effective and, indeed, pose many limitations. Nevertheless, each region's RDP specifies detailed criteria for allocating resources and assigning target areas. Landscape plans can be the source of criteria for "territorialising" rural policies (Cassatella and Seardo 2014; Rega 2014), ensuring more effective allocation of resources by designing ecological networks, reinforcing identity values and recreational potential. It is clear that, considering their environmental and agricultural value, these areas must be identified jointly by sectors and included in common databases.

Lastly, some of the mechanisms needed to place integration between landscape and agriculture on a workable basis exist but must be jump-started, while others must be created on the technical level. One possible route is to share knowledge, objectives and evaluation approaches between sectors at the level of regional policies. Hence, it is a matter of finding a common language, both in terms of factual knowledge (numerical and geographic data that characterise the landscape and relative policies), and in terms of the scale of representation (from the farm scale of the CAP to the landscape scale of the RLP). It also means integrating strategies to identify common areas of reference in order to channel the associated sectoral measures towards the rural landscapes where there is a real need. In addition, to be complete, integration must also contemplate the evaluative dimension, precisely the construction of shared evaluation frameworks that can enhance the monitoring system's efficiency and effectiveness, and take the multitude of values represented by the rural landscape into consideration.

Stating the values of rural landscape, which have so far only been celebrated or hazily mentioned by the CAP, will be presumably one of the new and irreconcilable challenges that will be faced over the coming decades, especially to maintain and improve mankind's well-being and quality of life.

Acknowledgements Section 7.2.2 is the result of a work supported by the ISI Foundation, CRT Foundation and IRES Piemonte of Torino, Italy (Lagrange scholarship, 2014–15). While Sects. 7.2.1 and 7.2.3 are the output of some research carried out at IRES Piemonte in the context of the evaluation activities of the Piedmont RDP 2014–2020. In addition to Dr. Stefano Aimone and the RDP evaluation staff of IRES, we would like to thank for the methodological support and the data provided: Mario Perosino (Piedmont Regional Authority), Filomena Nardacchione, Michela Sigliano, Lorena Cora and Emilio De Palma (CSI Piemonte), and Nicoletta Alliani (IPLA Piemonte).

References

Agnoletti, M. (Ed.). (2013). *Italian historical rural landscapes. Cultural values for the environment and rural development*. Dordrecth, Heidelberg, London, New York: Springer.

Cassatella, C., & Seardo, B. M. (2014). In search for multifunctionality: The contribution of scenic landscape assessment. In C. Rega (Ed.), *Landscape planning and rural development. Key issues and options towards integration* (pp. 41–60). Cham, Heidelberg, New York, Dordrecht, London: Springer.

Cassatella, C. (2015). Politiche rurali e politiche del paesaggio: un'alleanza (forse) possibile. *Proceeding of conference "XVII Conferenza SIU"*, Venezia, 11–13 June.

Cassatella, C., Gottero, E. (Eds.). (2016). Prove tecniche di alleanza tra politiche rurali e politiche del paesaggio. *Urbanistica Informazioni, 269–270*, 11–23.

Cevasco, R. (2013). Piedmont. In M. Agnoletti (Ed.), *Italian historical rural landscapes. Cultural values for the environment and rural development* (pp. 175–198). Dordrecth, Heidelberg, London, New York: Springer.

Commission of the European Communities (CEC). (1997). *Agenda 2000: For a stronger and wider union* (Vol. I), COM(97) 2000 final, Brussels. Accessed on 8 Mar, 2018 http://eur-lex.europa.eu/resource.html?uri=cellar:80958a30-795a-4152-99a5-cf86f455a211.0008.01/DOC_1&format=PDF.

European Court of Auditors (ECA). (2011). *Is agri-environment support well designed and managed?* (Special Report No 7/2011). Luxembourg: Publications office of the European Union. Accessed 9 Jan, 2015 http://www.eca.europa.eu/Lists/ECADocuments/SR11_07/SR11_07_IT.PDF.

Fanfani, D. (2013). Empowerment rurale e rigenerazione paesaggistico-ambientale del territorio agricolo. Una domanda di mobilitazione sociale e 'pianificazione dal basso. In D. Poli (Ed.), *Agricoltura paesaggistica. Visioni, metodi, esperienze* (pp. 179–196). Firenze: Firenze University Press.

Gottero, E. (2016a). *Un sistema complesso da valutare: il paesaggio rurale. Indicatori a sostegno delle politiche* [A complex system to be evaluated: the rural landscape. Indicators to support decision making]. Torino: IRES Piemonte. Accessed 8 Mar, 2018 http://www.digibess.it/fedora/repository/openbess:TO082-01931.

Gottero, E. (2016b). *Agricoltura Metropolitana. Politiche, pratiche e opportunità per l'innovazione territoriale nel torinese* [Metropolitan Agriculture. Policies, practices and opportunities for territorial innovation in Turin], Torino: IRES Piemonte. Accessed 8 Mar, 2018 http://www.digibess.it/fedora/repository/openbess:TO082-01941.

Gottero, E., & Cassatella, C. (2017). Landscape indicators for rural development policies. Application of a core set in the case study of Piedmont region. *Environmental Impact Assessment Review, 65,* 75–85. https://doi.org/10.1016/j.eiar.2017.04.002.

Jeanneaux, P., Aznar, O., & Déprés, C. (2011). Les services environnementaux fournis par l'agriculture et leurs modes de gouvernance: Un cadre d'analyse économique. *Regions and Cohesion, 1*(3), 117–144.

Lohrberg, F., Licka, L., Scazzosi, L., & Timpe, A. (Eds.). (2016). *Urban agriculture Europe*. Berlin: Jovis Verlag.

Piedmont Regional Authority. (2009). *Regional landscape plan* (RLP). Accessed 1 Dec, 2014 http://www.regione.piemonte.it/territorio/pianifica/ppr.htm.

Piedmont Regional Authority. (2012). *Rural development program RDP 2007–2013*. Version 19/12/2012. Accessed 1 July, 2013 http://www.regione.piemonte.it/agri/psr2007_13/misure/versioni.htm.

Piedmont Regional Authority. (2015a). *Rural Development Program 2014–2020* (RDP). Version 1.2 30/10/2015. Accessed 8 Mar, 2018 http://www.regione.piemonte.it/agri/psr2014_20/dwd/testoNegoziato/propDef/01_testo.pdf.

Piedmont Regional Authority. (2015b). *Regional landscape plan* (RLP). Accessed 8 Mar, 2018 http://www.regione.piemonte.it/territorio/pianifica/ppr.htm.

Piorr, A., Ungaro, F., Ciancaglini, A., Happe, K., Sahrbacher, A., Sattler, C., et al. (2009). Integrated assessment of future CAP policies: Land use changes, spatial patterns and targeting. *Environmental Science and Policy, 12,* 1112–1136.

Prager, K., Reed, M., & Scott, A. (2012). Encouraging collaboration for the provision of ecosystem services at a landscape scale-rethinking agri-environmental payments. *Land Use Policy, 29,* 244–249.

Rega, C. (Ed.). (2014). *Landscape planning and rural development. Key issues and options towards integration*. Cham, Heidelberg, New York, Dordrecht, London: Springer.

Rovai, M., Gorelli, S., & Santucci, A. (2008). Un modello di supporto alle decisioni per la definizione di politiche per il paesaggio rurale. *Agribusiness Paesaggio & Ambiente, XI*(3), 207–216.

Scolozzi, R., Santolini, R., & Morri, E. (2012). Delphi-based change assessment in ecosystem services values to support strategic spatial planning in Italian landscapes. *Ecological Indicators, 21,* 134–144.

Spaziante, A., Rega, C., & Carbone, M. (2013). Spatial analysis of agri-environmental measures for the SEA of rural development programmes. *Scienze regionali, 12*(2), 93–116.

Termorshuizen, J. W., Opdam, A. E. P., & Opdam, P. (2009). Landscape services as a bridge between landscape ecology and sustainable development. *Landscape Ecology, 24,* 1037–1052.

Vanni, F. (2013). Il possibile impatto dell'applicazione del greening in Italia. *Agriregionieuropa, Anno, 9*(35), 29–32. Accessed 23 Jan, 2014 http://www.agriregionieuropa.univpm.it/riviste/agriregionieuropa_n35.pdf.

Willemen, L., Verburg, P. H., Hein, L., & van Mensvoort, M. E. F. (2008). Spatial characterization of landscape functions. *Landscape and Urban Planning, 88,* 34–43.

Enrico Gottero (Editor) is lecturer in Landscape and Regional Planning at the Interuniversity Department of Regional and Urban Studies and Planning (Politecnico di Torino), as well as researcher at the Institute of Economic and Social Research of Piedmont (IRES). His main interests focus on the interaction between agriculture and rural-agrarian landscapes, urban-rural linkages, farmland preservation, urban agriculture, agri-food policies and practices. He is also author of several publications on these topics, including at international level.

Claudia Cassatella is an Architect, Ph.D. in Landscape Design/Planning and Associate Professor in Landscape and Regional Planning, Vice-Chair of the Faculty of Planning and Design at Interuniversity Department of Urban and Regional Studies and Planning (Politecnico di Torino). Her research focus on landscape, cultural and natural heritage, at different scale and aimed to different operational areas (landscape plans, strategic plans, management plans of protected areas and UNESCO sites, urban planning, rural development plans and related evaluation tools). She is member of the executive board of UNISCAPE Network of the Universities for the implementation of the European Landscape Convention. She is also author and editor of more than a hundred publications, among which Landscape Indicators (Springer, 2011).

Chapter 8
Foodscape and Food Urbanism in Europe: The Urban-Rural Interface

Susan Parham

8.1 Introduction

The central contention of this chapter is that the urban edges and regions surrounding European cities offer a dynamic set of foodscapes with critical urbanism implications. Some of food's most complex interplay with spatiality happens at the urban edge, however that edge is defined, itself a subject of contention. As urban conurbations intrude into traditional agricultural foodspace, and reshape socio-spatial practices in relation to food, maintaining and retrofitting peri-urban (and wider regional) areas towards sustainable urbanism becomes an increasingly challenging and necessary task for conceptualising and undertaking placemaking (Marsden 2013). This is the case for spatial design and for the planning and governance of place, for which it is argued there is a need to build a new food paradigm (Marsden and Morley 2014).

In this chapter primary research examples are drawn from a range of European edge urbanism conditions in food terms, although material from wider regions is also explored where pertinent to the theme. Fieldwork findings are from new town food retrofitting in the United Kingdom (north of London) and the revival of gastronomic landscapes on urban edges in Italy (around Florence) and France (around Montpelier). This primary data is framed and supported by a secondary review of research and applied practice from European peri-urban and rural contexts to delve into designing and planning urban edge food and urbanism territory. The chapter concludes with suggestions for the future to underpin and extend European peri-urban (and to an extent) regional foodscapes as sustainable urbanism.

S. Parham (✉)
Environment and Agriculture Group, UH Geography, University of Hertfordshire, Hatfield, England
e-mail: s.parham@herts.ac.uk

E. Gottero (ed.), *Agrourbanism*, GeoJournal Library 124,
https://doi.org/10.1007/978-3-319-95576-6_8

8.2 Framing the Chapter

What are the urban edges at issue here? The nature and extent of the peri-urban has been subject to a great deal of consideration in geography and cognate disciplines over the last fifty years. While urban morphologists have traced changing land use over time on the urban edge (Conzen 1960), defining exactly what the urban edge constitutes, or even what it should be called, has proved difficult given its complexity (Qviström 2007; Buxton and Choy 2007). Theorists have asserted the claims of the peri-urban 'borderlands' to being distinct in character rather than just in between the city and the countryside (Stilgoe 1998; Bunker 2002; Boume et al. 2003; Houston 2005; Simon et al. 2006; Madsen et al. 2010; Gonçalves et al. 2017). In part, the contested nature of the urban edge reflects the changing nature of urbanism itself as many cities have rapidly grown in spatial extent while others have shrunk (Neill and Schlappa 2016); and the forms that growth has taken have altered physically, socially and economically.

At the same time, theorists caution against treating this peri-urban zone as only requiring analysis of conversion of agricultural land to urban sprawl (Audirac 1999), much as problematising sprawl urbanism remains a legitimate area for research and practice (Frumkin et al. 2004; Phelps et al. 2006; Couch et al. 2007; Hennig et al. 2015; Richardson and Chang-Hee 2017). The conceptual challenges of the fast transforming, dynamic and jumbled urban region has seen it described in a wide range of syllogisms including exurbia, edge cities, superburbia, technoburb and megalopolis (Lapping and Furuseth 1999; Bontje and Burdack 2005; Buxton and Choy 2007).

While the intrusion of the urban is a fascinating aspect of the development of the urban edge and wider regions around cities, in this chapter the predominant focus is on the food side of the equation. Turning to food as part of this spatial and economic zone, theorists argue for recognition of the very place-specific foodspace implications of both the urban edge and rural spaces (Losada et al. 1998; Briquel and Collicard 2005; Trefon 2009; Opitz et al. 2016; Roe et al. 2016). There have been attempts over the long term to conceptualise and respond effectively to the interplay between rural and peri-urban food production and urban consumption, including from the work of Patrick Geddes (Meller 1994; Talen 2002) and the 1920s bioregionalism of the Regional Planning Association of America (Church 2014). Most recently attention has been drawn to such foodspace through alternative food geographies and networks (Maye et al. 2007; Jarosz 2008) and metropolitan regional food policy (Pothukuchi 2009), while more design based approaches are explored later in the chapter.

An era of multifunctional agriculture in rural regions and on the edge of cities at the rural-urban interface (Zasada 2011; Zasada et al. 2011) is matched by a burgeoning interest in urban agriculture (Brown and Jameton 2000). Theories of alternative food geographies and networks, shorter supply chains and localised food systems are helping to conceptualise and reframe how edge food space can be shaped or retrofitted along more sustainable lines than is apparent within mainstream food production and related arrangements (Parrot et al. 2002; Goodman 2004; Ilbery and Maye 2005; Maye et al. 2007; McClintock 2010; Paül and McKenzie 2013; Kneafsey et al. 2013).

The concept of the foodshed is relevant here. Defined as «*the geographical area from which a population derives its food supply*» (Peters et al. 2009, p. 2), the foodshed is grounded in attempts to give spatial and resilience form to the way city and country interconnect in relation to food (Hedden 1929; Getz 1991; Kloppenburg et al. 1996; Feagan 2007; Horst and Gaolach 2015). There is a range of interesting contemporary place-based work on how sustainable foodsheds operate (and can be supported) in different urban-rural context (Mougeot 1994; Tegegne et al. 2006; Friedmann 2014; Lengnick et al. 2015; Zasada et al. 2017).

Theorists are developing GIS based techniques to understand just how much urban edge agricultural land is actually used for agriculture (Verhoeve et al. 2015). Although such space is not expected to be able to provide for all of a city's food needs (Anglade et al. 2016), it is argued there is plenty of scope for 'recoupling' production and consumption in peri-urban areas, as for example around Paris (Tedesco et al. 2017). Such work can be supported by techniques including landscape plans and audits (Selman 2002; Aalders and Morrice 2012) and natural capital and ecosystems services approaches to creating, maintaining and retrofitting sustainable food systems (Elson et al. 2010; Bommarco et al. 2013).

There is also a theoretical and applied interplay with urban design based approaches with a focus on sustainable urbanism (Farr 2008) and sprawl avoidance or repair (Talen 2002; Dunham-Jones and Wlliamson 2009; Tachieva 2011). Here the conceptual tradition stemming from regional landscape characterisation and design is critical (Geddes 1915; McHarg 1969; Neuman 2000), including the transect and more latterly the food transect (Duany 2002; Bohl and Plater-Zyberk 2006). As Duany and DPZ (2011, p. 258) have noted, ' «*The transect is a fractal that allows design to integrate across scales: from regional tiers, to community codes, to architectural standards. At the largest scale, that of the region, the transect geographically allocates urban-to-rural tiers to accept varying degrees of environmental protection and development*»'.

Using a transect-based approach, some designers seek to shape space from urban to rural according to sustainable urbanism principles, with a central emphasis on food (Duany 2002). The food transect shows how, in each transect defined condition of urban space, a range of food possibilities can be built into place design and use over time, from a limited range of window boxes and roof gardens in the most urban conditions to more widely mixed and spatially extensive food-related land uses at the urban edge (Duany and DPZ 2011). Similarly food as a planning strategy and policy matter that connects to the good governance of place has emerged over the last twenty-five years (Parham 1992, 1993a; Pothukuchi and Kaufman 2000; Pothukuchi 2009; Morgan 2009; Wiskerke 2009; Donovan et al. 2011). Although, as Vitiello and Brinkley (2014) note, it has a much longer 'hidden' history, this comparatively more recent metropolitan regional planning emphasis suggests that increasingly there is acknowledgement that foodspace requires active planning in support. Moreover, this is an issue that ties urban and rural together as it did in and around traditional cities to which we now turn.

8.3 Historic and Gastronomic Context

Economic historians have noted the long-term interplay between urban and rural land uses in traditional urbanism, with productive space always sprinkled through and found at the edges of towns in both European and other contexts (Braudel 1988; Björklund 2010). Mok et al. (2014), for example, describe fascinatingly the very long-term interplay between urban and rural land uses in Japanese cities and peri-urban areas. Place was understood as critical to what food was produced, as in the Lea Valley cucumbers of London, or the vegetables of Edo districts in Japan which reflect where they were grown, such as "Nerima-daikon" (radish), "Komatsu-na" (rape), and "Yanaka-shouga" (ginger) (Mok et al. 2014, quoting Yokohari et al. 2010).

More broadly, urban-rural symbiosis has driven urban development through the creation of surpluses, benefitted urban dwellers and agricultural producers, and fashioneded very long term food landscapes around many European cities and towns (Kostof 1992, 1991; Parham 2015). These edges have had a particularly critical role to play in ensuring sufficient urban food provisioning can be found within walkable range of the urban centre (Kostof 1992). Peri-urban space has for millennia been used for pasturing animals, dairying, fruit orchards, vegetable gardens, viticulture and cropping among other productive land uses: it has also provided space for food processing and manufacture, storage, sale and consumption (Kostof 1992).

Turning to the modernist era, Ebenezer Howard's late 19th century model for garden cities reflected this requirement, putting food space and a spatially, economically and environmentally based local food system, at the heart of his proposals for a constellation of garden city settlements (Howard 1898; Hall and Ward 1998; Parham 2016; Parham and Abelman 2018). Food spaces were interwoven through the town but also formed much of the basis for a productive greenbelt that formed a growth boundary around each settlement (Howard 1898; Parham 2015). The post *Second World War* new towns created around Europe and elsewhere tried to build foodspace into a more limited extent (Alexander 2009), such as mid-block allotment gardens often later to fall into disuse and disrepair. In the United Kingdom at least, new towns were arguably less successful in integrating foodscapes into their urbanism than had been the garden cities that preceded and influenced them. More broadly, suburban areas in the 20th century global north continued to support some food production despite a modernist shift to consumption in domestic garden space particularly in the latter half of the century (Gaynor 2006; Parham 2015).

In the global south, meanwhile, direct spatial, economic and ecological foodspace connections at the urban edge have remained in place to varying extents (Marsh 1998; Laquian 2005; Redwood 2009). Yet even within such predominantly traditional urbanism urban edge productivity is under threat (Cheung 2007). In a European context, which is the focus of this chapter, the loss of productive and related land uses making up such foodspace is further along a path of disconnection from ecological and resilience realities. There are some exceptions to this predominant situation of foodspace loss such as a revival in a 19th century tradition of allotments which can be found across a number of European urban edges under different names (Parham 2015). Other examples are notable for their uniqueness, such as the *hortillonages* abutting Amiens which have been given special patrimony status through UNESCO (Jones 1997; Parham 2015). The socio-spatial practices associated with such foodspace in consumption terms, such as the *guinguettes* of peri-urban Paris (Brennan 1984) or the pleasure gardens of Vauxhall on the then edge of London that gave rise to the European 'Wauxhalls' (Conlin 2008), are of continued but now largely historical interest.

Gastronomers have decried the loss of connectivity as urban development on the edges of European towns and cities replaced and destroyed productive rural uses at urban edges (Grigson 1978; Parham 1992, 2015). Especially from the latter part of the 20th century onwards, the nature of European cities shifted towards sprawling forms of urban development (Munoz 2003; Catalán et al. 2008; Salvati and Sabbi 2011). These development forms (often referred to by the catch-all term 'sprawl') have taken up land previously used for forms of agricultural production and related land uses (Couch et al. 2007; Parham 2015).

From the rural side, a presumption of primacy has emerged by which agricultural producers and other landholders have assumed that urban development is likely to replace their food growing activities (Bunker and Houston 2003; Cook and Harder 2013). Accepting they are farming «*suburbs in waiting*» (Bunker and Holloway 2001, p. 13) and believing they have 'development rights' to sell farm land at urban development prices, farmers have sought to benefit financially from what is seen as an inevitable change in land use (Bunker and Houston 1992, p. 24). On the urban growth side of the equation, rising incomes and a spatial range predicated on car access has widened the territory within which city dwellers feel able to settle, sometimes to embark on forms of 'sustainable' small-scale farming themselves (Beilin et al. 2015).

8.4 Urban Edges as Foodspace Today

If the urban edge has transformed to predominantly non-productive space in the city's countryside (Bryant and Johnston 1992) this has not been a simple causative relationship of shift from rural to urban (Murdoch and Marsden 2013; Parham 2015). Rather, complex shifts in the interplay between productive and unproductive space, and transforming socio-spatial practices in relation to food and urbanism, are evident on European cities' and towns' urban edges (Brueckner 2001; Boume et al. 2003). A cost-price squeeze in farming has led to both higher levels of productivity and farm restructuring and diversification into non-farm activities as part of 'post productivism' (Bunker and Houston 1992; Tacoli 1998; Lanjouw and Lanjou 2001; Briquel and Collicard 2005; Reardon et al. 2007; Preasthom and Kristensen 2007). Developing land use multifunctionality (Zasada et al. 2011) and other trends including rural gentrification and counter-urbanisation are also in evidence (Butt 2014).

Yet edge of town locations in Europe are often best suited to highly intensive, high-value agriculture, horticulture and viticulture (Opitz et al. 2016) and can contribute to shorter food supply chains (Aubry and Kebir 2013) and well as other resilience benefits. It is at urban edges that new products and markets which underpin farm activities may be developed, and their economic value can sometimes in turn protect these landscapes from urban redevelopment (Parham 1996, 2015, 2012). The resultant sustainability and conviviality benefits related to gastronomy are seen as critical to resilient futures (Miele and Murdoch 2002; Regoli et al. 2011; Parham 2014; De Zeeuw and Dreschsel 2015).

The sustainability implications of relatively unconstrained urban growth are increasingly understood at a theoretical level (Hough 1984; Rees 1992; Beatley 2000; Tjallingi 2000; Wheeler and Beatley 2014). As noted in Parham (2015), concepts and tools such as ecological footprinting, ecosystem services and urban metabolism modelling help conceptualise, measure and respond to the impacts including in relation to food (Allen 2003; Grimm et al. 2008; Wiskerke and Viljoen 2012). These are more and more important as cities absorb food and other resources from areas far beyond their boundaries, and create carbon and other effects which impact widely (Wackernagel and Rees 1996; Giradet 1999; Parham 2015, p. 201). Allen (2003, p. 140) has pointed out the particularly pernicious effects on peri-urban areas of expanding urban footprints.

Attitudinal surveys suggest urbanites value peri-urban areas as multifunctional landscape areas they do not wish to see solely developed for housing (Ivesa and Kendal 2013). Yet, in practice, housing and other urban land uses are replacing farmland at a huge rate, while agriculture is permeated with assumptions that proximity to the places where food is consumed is no longer required. In both cases it is argued that there is a lack of applied acknowledgement of, or sufficient response to, the negative environmental effects of dominant practices in urbanism and in food production (Hough 1984; Flannery 1994; Ilbery et al. 1997; Allen 2003; Chen 2007; Wiskerke and Viljoen 2012). This matters not just in gastronomic landscape terms (Hardy 1993, 1994) but for food resilience, especially given sharpening climate change effects (Aguilar et al. 2003; Paul and Tonts 2005; Torres et al. 2007).

Today the urban edges of European cities and towns are critical sites for the growing phenomenon of gastronomic tourism (Parham 1995; Hjalager and Corigliano 2000; Hjalager and Richards 2004; Urry and Larsen 2011; Sidali et al. 2011; Cánoves and de Morais 2011; Long 2012; Parasecoli and de Abreu e Lima 2012). Sustainable gastronomy may be seen as a tourism product (Scarpato 2002; Hjalager 2002: 21) but its spatial and design dimensions may have been underplayed (Parham 2015). Richards (2015) has argued that there has been a shift from gastronomy as a rather 'themed' aspect of the experience economy to one where gastronomically focused tourism experiences are co-created by producers and consumers, and finally to a richer situation in which communities are developed around gastronomy. Richards (2015) situates this as a shift from foodies to foodscapes.

Framed in this way the peri-urban can be seen as increasingly important, as it offers a synchronicity between tourists and artisan agriculture and other supposedly niche food land uses (Boniface 2003). Urban edges can be particularly rich territory for gastronomically informed restaurants and other foodspaces such as wineries, mills, farm shops, nurseries, apiaries, markets, artisan product production spaces, wine caves, tasting rooms, and accommodation, sometimes in an *agritourismo,* ecotourism, or slow tourism mode (Ignatov and Smith 2006; Fullagar et al. 2012; Parham 2015, p. 208).

The evolution of such foodspaces reflects both changing socio-spatial food practices and relocalising food efforts occurring in a wide variety of places (reported in Parham 2015, p. 249), bringing together new kinds of producer groupings, trade circuits and shorter supply chains (Renting et al. 2003; Sonnino and Marsden 2006; Fort et al. 2007). As part of a gastronomic imaginary, these peri-urban spaces are the subject for cookbooks, television programmes, online accounts, and magazine and newspaper articles about particular chefs' recipes and restaurants; a sign of the 'growing virtual life of the peri-urban in the gastronomic imagination' (Parham 2015, p. 208; Scott and Duncan 2016; Frost et al. 2016).

This material and virtual positioning has attracted criticism in Foucauldian terms as marginalising working class foods (de Jong and Varley 2017), and seen as playing out of conspicuous consumption and tastefulness in Bourdieu's sense of establishing an individual habitus that may travel from home with the individual (Lee et al. 2014). It is also criticised for potentially '*erasing untidy or less appealing contexts such as sprawl housing, autoroutes or distribution landscapes in the vicinity'* (Parham 2015, p. 208). Thus, within overall arguments that localising food systems is required at the urban edge (and the broader region), gastronomic tourism can be situated as reinforcing or at least failing to address inequality. It follows that proposals and practices aimed at re-regionalism of food systems may fail to deal effectively with social and spatial inequalities (Donald et al. 2010; Goodman et al. 2010).

Yet, insofar as this discourse deals with spatial aspects, it tends to assume all food-focused land uses are at the luxury end of the tourism market. It underplays considerable gastronomic landscape and sustainable economy diversity and the benefits that a broad range of different foodspaces underpins. Moreover, a range of peri-urban and wider regional foodspaces that are neither exclusive nor expensive are challenging global commodity chain norms. These are increasingly evident along the food chain, from production through processing and distribution to consumption (Renting et al. 2003). Through such foodscapes, place is seen as a basis for food traditions, authenticity and quality (Bessière 1998; Feagan 2007; Sims 2009).

Part of the basis for asserting the primacy of place comes from its links to food's *terroir* and appellation based methods and places of production. Defined as «*an area or terrain, usually rather small, whose soil and micro-climate impart distinctive qualities to food products*» (Barham 2003, p. 131) *terroir* is famously reflected in products like wine and cheese but other products (like Breton oysters and *sarrasin* [buckwheat], or the *lardo* and *cannellini* of Tuscany) are also products which come from particular gastronomic landscapes (Roux and Roux 1989; Barkat and Vermignon 2006).

Various forms of geographical identification of specific foods and food products have been developed across a range of European countries (Bessière 1998: 25). These indicators of local character and quality include geographical indications (GIs) and other placed based designations, like *appellation controlee*, *Appellation d'Origine Contrôlée* (AOC), *l'appellation Montagne*, DOCG, *Protected Denomination of Origin* (PDO), and *Protected Geographical Indication* (PGI) (Parham 2015, p. 259). As Meiberger and Weichbold (2010, p. 1626) note: «*Protected status differentiates and reflects the fragility of such agricultural systems and products in the face of food commercialisation*». These in turn support «*small-scale economic health, gastronomic landscape preservation, ecology and regional food tourism*» (Parham 2015, p. 260). Famous examples of this encompass the area in central Northern Italy which *parmagiano reggiano* is produced (de Roest and Menghi 2000, p. 440).

Globally there is an intense fight underway between such place-based approaches and the delocalising thrust of *World Trade Organisation* rules and international trade negotiations including the *Transatlantic Trade and Investment Partnership* (TTIP) and the *Trans-Pacific Partnership* (TTP). In this area a sharp difference exists between a European focus on regionality and ethical production contrasting with technocratic and placeless American perspectives (van Otterloo 2005, p. 259). That is not to say there is unanimity within Europe where issues of regional and national food distinctiveness remain in play (Castellanos and Bergstresser 2006). Yet in places where industrialised agriculture has become completely dominant, the notion of retrofitting *terroir* has emerged as a valuable way of reverse engineering agrarian, social, gastronomic and environmental stewardship of place towards more sustainable, localised ends (Paxson 2010, p. 444, in Parham 2015; Gyimóthy and Mykletun 2017).

The *Slow Food* movement, founded in Piedmont in 1986, has sought to assert place-based approaches to food, with an emphasis on territorial patrimony, quality and conviviality (Petrini et al. 2005; Mayer and Knox 2006). *Slow Food* has been criticised on the grounds that some individual tourists incorporate its narratives into their elite tourism experience (Chrzan 2006), and that it has arguably been deradicalised as invoked in some American circumstances (Paxson 2005). However, this argued appropriation does not undermine its broader 'in principle' and practice claims as a movement. *Slow Food*, and the related *Città Slow* established in 1999, can be argued to exemplify a number of the urbanism points being made in this chapter through a focus on 'practical aesthetics' covering local distinctiveness in cuisine, materials, products, planning and place design (Miele and Murdoch 2002: 318). For example, to join the *Città Slow network* (which in 2017 covered 236 cities in 30 countries), cities must meet a range of requirements set out in the association's

Charter, which cover design, planning and governance for food-focused urbanism in towns, on urban edges and in surrounding regions.

8.5 Two Examples from European Applied Research

8.5.1 Retrofitting for Food in Hatfield, Hertfordshire

This theory based discussion offer a context for exploring examples from practice in the United Kingdom, France and Italy to consider aspects of the urbanism at play in urban edge and regional foodscapes. The first of these is a recent transect-based food-retrofitting example from primary research by the author in and around the edge of Hatfield, a new town in the county of Hertfordshire, just north of London in the United Kingdom. In 'edge' terms, new towns are themselves often located at the perimeters of major cities, as in this case close to London. Social, economic, environmental and spatial difficulties have emerged over time, in part associated with their modernist design, including expensive to maintain and underused green spaces (Alexander 2009).

The town of Hatfield is argued to be undergoing processes of spatial and economic shrinkage but also elements of growth at the same time. Some are experiencing the town as an expanding, economically advantageous space while others are suffering the ongoing shrinkage effects of de-industrialisation (Parham and McCabe 2016; Parham 2016). Initial analysis of conditions in and around Hatfield suggested that the new town offers a number of opportunities for food-centred retrofitting including on its own edges, such as orchards, allotments, gardens, and markets, in keeping with gastronomic strategy (Parham 1992), convivial green space (Parham 1993b) and food transect ideas noted earlier in the chapter.

The researchers sought to understand how foodspace interventions based on sprawl repair techniques could help underpin more sustainable urbanism in the new town in future within a range of different transect 'continuum' conditions from town centre through to peri-urban rural space (Dunham-Jones and Williamson 2009; Tachieva 2011; Parham and Abelman 2018). A series of design scenarios were developed for retrofitting existing space that was argued to be dysfunctional or suboptimal in food and urbanism terms. The design approaches made particular reference to European based repair techniques (Vall-Casas et al. 2011). These included the potential to assist those being excluded from affordable, good quality food access and related food resilience benefits, as identified in the town through baseline research (Parham and McCabe 2016; Parham 2016).

To contextualise, the edge of Hatfield New Town has itself undergone a number of design interventions since first being developed in the post war era. Following closure of the large former aerospace industry site located on the north-western edge of Hatfield, a substantial spatial area was remade into a university campus and business park in the international campus-garden-suburb style (Forsyth and Crewe 2010).

This model has been in some ways an economic success story, albeit with largely unacknowledged sustainability externalities caused by its predominantly mono-functional land take, designed-in car dependency, and large-scale, sterile landscape setting. The business park, in particular, contains a large amount of green space but one that only offers this in a very passive form of 'privately owned public space' (Schmidt et al. 2011). The manicured, access and use controlled landscapes abutting object buildings, very big car parks and major roads are primarily for viewing from offices. Sign boards clearly state its status as private space, and that all ball games, picnics or other uses are forbidden. The other substantial open space is of greyfield tracts (former aerodrome land) still awaiting back office and large shed type developments. These spaces are used for informal pedestrian cut-throughs to make marginally more walkable the vehicle-oriented superblock road layout. It was argued that despite its economic success, the business park's townscape related paucity of food options (production and consumption) meant that it constituted a kind of rich food desert.

One of the proposed retrofits responded directly to this problematic food context with proposals for a transitional retrofitting scenario. In work published in 2016 it was proposed that both these landscapes, and the broader greyfield areas, could be planted productively, including with allotments, community gardens and orchards where appropriate (Fig. 8.1) (Parham and McCabe 2016). These would be managed but also offer community access and foraging opportunities. Given the sheer scale of the spaces it was suggested that an ornamental *potager* garden or gardens in the style of Bosmelet in Normandy, where vegetables, fruits and herbs are both ornamental and picked for use, might be considered as part of the forageable landscape design.

This retrofitting proposal was offered as a short-to-medium term intervention, prior to the more desirable outcome of a complete remaking of the business park as sustainable, walkable, food-centred urbanism fully embracing sprawl repair principles as they apply to edge of town and urban boundary areas (Tachieva 2011). Other edge space food retrofitting scenarios proposed included for orchards near housing areas at the perimeter of the new town and a *potager* garden in a local hamlet, Mill Green, on Hatfield's fringe (Fig. 8.2).

8.5.2 Rural-Urban Edge, Montpelier and Florence

The second example is earlier research on gastronomic landscapes in the vicinity of Montpelier and Florence with a focus on the revival of small-scale viticulture, high value wine making and renewal of farm buildings by urban professionals using their city derived incomes to finance these activities (Parham 1995). Although completed in the 1990s, more recent research about the area around Pisa has reinforced some conclusions emerging from the original work (Orsini 2011, 2013) so it is argued to be worth briefly touching on salient aspects here.

Fig. 8.1 Business park food retrofit *Source* Parham and McCabe (2016), illustration by Ben McCabe

In both Montpelier and Florence the urban edge had been an important food and wine production area, that had shaped the urban region over hundreds of years. In Italy this landscape stability of landed estates, *fattoria* and effectively indentured peasantry was somewhat formalised through the (very exploitative) *mezzadria* system (Roden 1989; Casanova and Memoli 2010). In both Italy and France, change away from traditional agriculture since the 1950s saw the landscapes around Florence and Montpelier subject to considerable alteration and disruption, and vineyards, olive orchards and farm buildings were abandoned. In the late 20th century the precarious incomes of upland farmers were further undermined by deregulatory farming regimes (Shucksmith and Ronningen 2011; Orsini 2013) yet have demonstrated considerable broader sustainability value (O'Rourke et al. 2016).

In both peri-urban Florence and Montpelier over the last few decades high quality edge-of-town food and drink production has been revived, in part by urban 'hobby' farmers with strong gastronomic sensibilities (Parham 2015). Thus, around Montpelier, university academics (for example) bought abandoned vineyards and brought these back into wine production. Similarly urban professionals in Florence purchased derelict *fattorias* and olive orchards near the urban edge, and started niche olive oil production, processing and related *agritourismo* and marketing (Parham 1995). As Orsini (2013) notes in relation to an analogous area around Pisa,

Fig. 8.2 Mill Green potager *Source* Parham and McCabe (2016), illustration by Ben McCabe

this approach is not just an instrumental economic one, but reflects a view of upland farms and farming as part of the region's cultural heritage. In both case studies «*food production and processing activities have helped revitalise the economic capacity of city peripheries as places of agricultural innovation, while maintaining or renewing natural environments and food spaces such as hillside terraces and abandoned wine caves*» (Parham 2015, p. 210).

In this example the urbanism of the urban edge has been renewed through food-focused production, tourism and promotional activity. In a positive spatial, social and economic feedback loop, increasing the landscape quality of the edge through value adding food production and tourism has, in turn, made the peri-urban area more economically viable as foodspace. As noted previously, «strong links have been developed between built and cultural heritage; between food production and gastronomic tourism; and from artisan producer to central city food market places» (Parham 2015, p. 210, quoting Parham 1995). In conclusion in relation to urbanism aspects of these peri-urban foodscapes, both the Montpelier and Florence examples melded farm and non-farm land uses and activity on urban edges to support regional economies and undercut sprawl. In both places, this *symbiotic urbanism relationship between urban and rural has played a key part in food-led revitalisation, innovation and conviviality on the urban edge* (Parham 2015).

8.6 The Chapter in Review

This chapter has focused on aspects of food urbanism in Europe in the context of the design and planning of urban-rural spaces, interfaces and linkages. It has posited that urbanism in relation to food is particularly complicated on the transforming edges of cities. The predominant conditions of sprawling growth of urban areas and transformations to food production and other food-related land uses, offer both threats and in some cases opportunities to retrofit in more food conscious design and planning. Examples shown in the chapter demonstrate a rich social and economic interplay between urban and peri-urban land uses.

The chapter has been structured to review existing place-making research and offer case studies of applied primary source research to delve into how such food-related urbanism is being represented in a series of peri-urban foodscapes. Techniques referenced in this chapter have been used to both identify food and urbanism issues in edge space, and then apply design and other interventions based on well-founded urbanism principles. These have been posited as ways to help improve food's sustainability and conviviality outcomes at the urban edge.

A number of tools and techniques, including to audit landscapes, establish ecological footprints, model urban metabolisms and support sustainable foodsheds are now available to help in identifying, measuring and understanding urban edge foodscapes. It has bee argued that sustainable design approaches, including retrofitting techniques based on transects and reverse engineering *terroir*, among others, can help in developing interventions to repair foodscapes under threat from issues such as sprawl on the one hand and environmentally damaging agricultural norms on the other. Although the focus in the chapter has predominantly been on urban design it is understood that some forms of gastronomic tourism, initiatives like Slow Food and Slow Cities, and strategic and land use planning policy, can all also help underpin more positive food-centred peri-urban and rural-urban interfaces.

In conclusion, the gastronomic landscapes (Hardy 1993, 1994) of the urban edge of European cities offer critical sites for creating, maintaining and retrieving sustainable food-focused urbanism in future. The chapter demonstrates that there are serious issues at play but there is also very considerable scope to reconfigure food's relationship to urbanism in peri-urban and rural-urban areas in ways which are of benefit to us all.

References

Aalders, I. & Morrice, J. (2012). Land Use change and ecosystem delivery by green networks. land management: potential, problems and stumbling blocks, *Land Management, 21*.

Aguilar, A. G., Ward, P. M., & Smith, C. B., Sr. (2003). Globalization, regional development, and mega-city expansion in Latin America: analyzing Mexico City's peri-urban hinterland. *Cities, 20*(1), 3–21.

Alexander, A. (2009). *Britain's new towns: Garden cities to sustainable communities*. New York: Routledge.

Allen, A. (2003). Environmental planning and management of the peri-urban interface: Perspectives on an emerging field. *Environment and Urbanization, 15*(1), 135–148.

Anglade J., Medina M. R., Billen, G. & Garnier, J. (2016). Organic market gardening around the Paris agglomeration: Agro-environmental performance and capacity to meet urban requirements. *Environmental Science and Pollution Research,* 1–10. https://doi.org/10.1007/s11356-016-6544-1.

Aubry, C., & Kebir, L. (2013). Shortening food supply chains: A means for maintaining agriculture close to urban areas? The case of the French metropolitan area of Paris. *Food Policy, 41,* 85–93.

Audirac, I. (1999). Unsettled views about the fringe: Rural-urban or urban-rural frontiers? In O. J. Furuseth & M. B. Lapping (Eds.), *Contested countryside: The rural urban fringe in North America* (pp. 7–32). Aldershot: Ashgate.

Barham, E. (2003). Translating terroir: The global challenge of French AOC labeling. *Journal of Rural Studies, 19*(1), 127–138.

Barkat, S. M., & Vermignon, V. (2006). *Gastronomy tourism: A comparative study of two french regions: Brittany and La Martinique*. Paper presented at the Sustainable Tourism with Special Reference to Islands and Small States Conference, Malta, 25–27 May 2006.

Beatley, T. (2000). *Green urbanism: Learning from European cities*. Island Press.

Beilin, R., Reichelt, N., & Sysak, T. (2015). Resilience in the transition landscapes of the peri-urban: From 'Where' with 'Whom' to 'What'. *Urban Studies, 52*(7), 1304–1320.

Bessière, J. (1998). Local development and heritage: Traditional food and cuisine as tourist attractions in rural areas. *Sociologia Ruralis, 38*, 21–34.

Björklund, A. (2010). *Historical urban agriculture: Food production and access to land in Swedish towns before 1900*, Stockholm University, Faculty of Social Sciences, Department of Human Geography.

Bohl, C. C., & Plater-Zyberk, E. (2006). Building community across the rural-to-urban transect. *Places, 18*(1), 4–17.

Bommarco, R., Kleijn, D., & Pottsm, S. G. (2013). Ecological intensification: Harnessing ecosystem services for food security. *Trends in Ecology & Evolution, 28*(4), 230–238.

Boniface, P. (2003). *Tasting tourism: Travelling for food and drink.* Aldershot: Ashgate.

Bontje, M., & Burdack, J. (2005). Edge cities, European-style: Examples from Paris and the randstad. *Cities, 22*(4), 317–330.

Boume, L. S., Bunce, M., Taylor, L., Luka, N., & Maurer, J. (2003). Contested ground: The dynamics of peri-urban growth in the Toronto region. *Canadian Journal of Regional Science l Revue Canadienne des Sciences Régionales, XXV, 1*(2&3), 251–270.

Braudel, F. (1988). *The identity of France* (Vol. 2). Harper Collins: People and Production.

Brennan, T. (1984). Beyond the barriers: Popular culture and Parisian guinguettes. *Eighteenth-Century Studies, 18*(2), 153–169.

Briquel, V., & Collicard, J. J. (2005). Diversity in the rural hinterlands of European cities. In K. Hoggart (Ed.), *The city's hinterland: dynamism and divergence in Europe's peri-urban territories* (pp. 19–40). Aldershot: Ashgate.

Brown, K. H., & Jameton, A. L. (2000). Public health implications of urban agriculture. *Journal of Public Health Policy, 21*(1), 20–39.

Brueckner, J. K. (2001). Urban sprawl: Lessons from urban economics. *Brookings-Wharton papers on urban affairs, 1,* 65–97.

Bryant, C. R., & Johnston, T. R. R. (1992). *Agriculture in the city's countryside*. London: Belhaven Press.

Bunker, R. (2002). In the shadow of the city: The fringe around the Australian metropolis in the 1950s. *Planning Perspectives, 17*(1), 61–82.

Bunker, R., & Holloway, D. (2001). Fringe city and contested countryside: Population trends and policy developments around Sydney. Issues Paper No. 6, Urban Frontiers Program: University of Western Sydney.

Bunker, R., & Houston, P. (1992). At and beyond the fringe: Planning around the Australian city with particular reference to Adelaide. *Urban Policy and Research, 10*(3), 23–32.

Bunker, R., & Houston, P. (2003). Prospects for the rural-urban fringe in Australia: Observations from a brief history of the landscapes around Sydney and Adelaide. *Australian Geographical Studies, 41*(3), 303–323.

Butt, A. (2014). Developing typology of changing multi-functional regions. *The Australasian Journal of Regional Studies, 20*(2), 233–257.

Buxton, M., & Choy Low, D. (2007). Change in peri-urban Australia: Implications for land use policies. In Hamnett, S. (Ed.), *Proceedings of the state of Australian cities national conference 2007* (pp. 291–302). Adelaide, Australia, 28–30 November.

Cánoves, G., & de Morais, R. S. (2011). New forms of tourism in Spain: Wine, gastronomic and rural tourism. In R. M. Torres & J. H. Momsen (Eds.), *Tourism and agriculture: New geographies of consumption*. Production and Rural Restructuring: Routledge.

Casanova, P., & Memoli, A. (2010). *La Romagna toscana. Mille anni di caccia, mezzadria e carbone.* Edizioni Sarnus.

Castellanos, E., & Bergstresser, S. M. (2006). Food fights at the EU table: The gastronomic assertion of Italian distinctiveness. *European Studies: A Journal of European Culture, History and Politics, 22*(1), 179–202.

Catalán, B., Saurí, D., & Serra, P. (2008). Urban sprawl in the Mediterranean?: Patterns of growth and change in the Barcelona Metropolitan Region 1993–2000. *Landscape and Urban Planning, 85*(3), 174–184.

Chen, J. (2007). Rapid urbanization in China: A real challenge to soil protection and food security. *CATENA, 69*(1), 1–15.

Cheung, S. C. H. (2007). Fish in the marsh. A case study of freshwater fish farming in Hong Kong. In Cheung, S., & Tan, C-B. (Eds.), *Food and foodways in Asia: Resource, tradition and cooking* (pp. 37–50). London: Routledge.

Chrzan, J. (2006). Why study culinary tourism? Answers for a healthy life: Research notes. *Expedition, 40*(1), 40–41.

Church, S. P. (2014). Exploring Urban Bioregionalism: a synthesis of literature on urban nature and sustainable patterns of urban living. *Sapiens, 7*(1), https://journals.openedition.org/sapiens/1691.

Conlin, J. (2008). Vauxhall on the boulevard: Pleasure gardens in London and Paris, 1764–1784. *Urban History, 35*(1), 24–47.

Conzen, M. R. G. (1960). Alnwick, Northumberland: A study in town-plan analysis. Transactions and Papers, *Institute of British Geographers* 27, iii + ix-xi + 1+3-122.

Cook, N., & Harder, S. (2013). By accident or design? Peri-urban planning and the protection of productive land on the urban fringe. In Q. Farmar-Bowers, V. Higgins, & J. Millar (Eds.), *Food security in Australia* (pp. 413–424). US: Challenges and Prospects for the Future Springer.

Couch, C., Leontidou, L., & Petschel-Held, G. (Eds.). (2007). *Urban sprawl in Europe*. Landscapes, Land-Use Change and Policy: Blackwell. RICS Research.

de Jong, A., & Varley, P. (2017). Food tourism policy: Deconstructing boundaries of taste and class. *Tourism Management, 60,* 212–222.

de Roest, K., & Menghi, A. (2000). Reconsidering `traditional' food: the case of Parmigiano Reggiano cheese. *Sociologia ruralis, 40*(4), 439–451.

De Zeeuw, H., & Dreschsel, P. (Eds.). (2015). *Cities and agriculture*. Routledge: Developing resilient urban food systems. New York.

Donald, B., Gertler, M., Gray, M., & Lobao, L. (2010). Re-regionalizing the food system? *Cambridge Journal of Regions, Economy and Society, 3*(2), 171–175.

Donovan, J., Larsen, K., & McWhinnie, J. A. (2011). *Food-sensitive planning and urban design: A conceptual framework for achieving a sustainable and healthy food system.* David Lock Associates, University of Melbourne and National Heart Foundation of Australia.

Duany, A. (2002). Introduction to the special issue: The transect. *Journal of Urban Design, 7*(3), 251–260.

Duany, A., & DPZ (2011). *Theory and practice of agricultural urbanism,* London, Duany Plater-Zyberk and Co. & The Prince's Foundation.
Dunham-Jones, E., & Williamson, J. (2009). *Retrofitting Suburbia: Urban design solutions for redesigning suburbs*. Hoboken: Wiley.
Elson, F., Foley, C., Foley, L. S., Leposo, A., Loure, E., Peterson, D., et al. (2010). Payments for ecosystem services as a framework for community-based conservation in Northern Tanzania. *Conservation Biology, 24,* 78–85.
Farr, D. (2008). *Sustainable urbanism*. Hobken, Wiley: Design with Nature.
Feagan, R. (2007). The place of food: mapping out the 'local' in local food systems. *Progress in Human Geography, 31*(1), 23–42.
Flannery, T. (1994). *The future eaters*. New York: George Braziller.
Forsyth, A., & Crewe, K. (2010). Suburban technopoles as places: The international campus-garden-suburb style. *Urban Design International, 15,* 165–182.
Fort, F., Ferras, N., & Amilien, V. (2007). Hyper-real territories and urban markets: changing conventions for local food–case studies from France and Norway. In Amilien, V., & Holt, G. (Eds.), *From local food to localised food. Special issue on local food products and systems. Anthropology of Food* S2: https://journals.openedition.org/aof/446.
Friedmann, H. (2014). Reflections on foodsheds in three continents. *Canadian Food Studies, 1*(1), 4–9.
Frost, W., Laing, J., Best, G., Williams, K., Strickland, P., & Lade, C. (2016). *Gastronomy, tourism and the media.* Channel View Publications.
Frumkin, H., Lawrence, F., & Jackson, R. J. (2004). *Urban sprawl and public health: Designing, planning, and building for healthy communities.* Island Press.
Fullagar, S., Markwell, K., & Wilson, E. (2012). *Slow tourism: Experiences and mobilities.* Bristol, Buffalo, Toronto: Channel View Publications.
Gaynor, A. (2006). *Harvest of the suburbs: An environmental history of growing food in Australian cities*. Crawley: WA, University of Western Australia Press.
Geddes, P. (1915). *Cities in Evolution. An introduction to the town planning movement and to the study of civics.* London: Williams and Norgate.
Getz, A. (1991). Urban foodsheds. *Permaculture Activist, 24,* 26–27.
Girardet, H. (1999). *Creating sustainable cities (No. 2).* Chelsea: Green Publishing.
Gonçalves, J., Gomesa, M. C., Ezequiela, S., Moreira, F., & Loupa-Ramosa, I. (2017). Differentiating peri-urban areas: A transdisciplinary approach towards a typology. *Land Use Policy, 63,* 331–341.
Goodman, D. (2004). Rural Europe redux? Reflections on alternative agro-food networks and paradigm change. *Sociologia Ruralis, 44*(1), 3–16.
Goodman, M. K., Goodman, D., & Redclift, M. (2010). Introduction: Situating consumption, space and place. In M. K. Goodman, D. Goodman, & M. Redclift (Eds.), *Consuming space: Placing consumption in perspective* (pp. 3–40). Aldershot: Ashgate.
Grigson, J. (1978). *The vegetable book.* Harmondsworth: Penguin Books.
Grimm, N. B., Faeth, S. H., Golubiewski, N. E., Redman, C. L., Wu, J., Bai, X., et al. (2008). Global change and the ecology of cities. *Science, 319*(5864), 756–760.
Gyimóthy, S., & Mykletun, R. J. (2017). The reinvention of terroir in Danish food place promotion. *European Planning Studies, 25*(7): Special Theme Issue: Nordic Food.
Hall, P., & Ward, C. (1998). *Sociable cities: The legacy of Ebenezer Howard*. Chichester: John Wiley and Sons.
Hardy, M. (1993). *The gastronomic landscape: Food production and the cultural value of countryside*. Paper presented in Gastronomic Symposiette series: University of Adelaide.
Hardy, M. (1994). *The future of food and Dining in Post-modern France*. Paper presented in Gastronomic Symposiette series: University of Adelaide.
Hedden, W. P. (1929). *How great cities are fed*. Boston, MA: D.C. Heath and Company.
Hennig, E. I., Schwick, C., Soukup, T., Orlitová, E., Kienast, F., & Jaeger, J. A. G. (2015). Multi-scale analysis of urban sprawl in Europe: Towards a European de-sprawling strategy. *Land Use Policy, 49,* 483–498.

Hjalager, A. M. (2002). A typology of gastronomy tourism. In Hjalager, A. M., & Richards, G. (Eds.), *Tourism and gastronomy* (pp. 21–35). Psychology Press.
Hjalager, A. M., & Corigliano, M. A. (2000). Food for tourists—Determinants of an image. *International Journal of Tourism Research., 2*(4), 281–293.
Hjalager, A. M., & Richards, G. (Eds.). (2004). *Tourism and gastronomy.* Psychology Press.
Horst, M., & Gaolach, B. (2015). The potential of local food systems in North America: A review of foodshed analyses. *Renewable Agriculture and Food Systems, 30*(5), 399–407.
Hough, M. (1984). *City form and natural process: Towards a new urban vernacular.* London, New York: Routledge.
Houston, P. (2005). Re-valuing the Fringe: Some findings on the value of agricultural production in Australia's Peri-Urban Regions. *Geographical Research, 43*(2), 209–223.
Howard, E. (1898). *To-Morrow: A Peaceful Path to Real Reform.* London.
Ignatov, E., & Smith, S. (2006). Segmenting canadian culinary tourists. *Current Issues in Tourism, 9*(3), 235.
Ilbery, B., Chiotti, Q., & Rickard, T. (Eds.). (1997). *Agricultural restructuring and sustainability: A geographical perspective,* CAB International.
Ilbery, B., & Maye, D. (2005). Food supply chains and sustainability: evidence from specialist food producers in the Scottish/English borders. *Land Use Policy, 22*(4), 331–344.
Ivesa, C. D., & Kendal, D. (2013). Values and attitudes of the urban public towards peri-urban agricultural land. *Land Use Policy, 34,* 80–90.
Jarosz, L. (2008). The city in the country: Growing alternative food networks in Metropolitan areas. *Journal of Rural Studies, 24*(3), 231–244.
Jones, L. (1997). *Kitchen gardens of France*. London: Thames and Hudson.
Kloppenburg, J., Jr., Hendrickson, J., & Stevenson, G. W. (1996). Coming into the foodshed. *Agriculture and Human Values, 13*(3), 33–42.
Kneafsey, M., Venn, L., Schmutz, U., Bálint, B., Trenchard, L., Eyden-Woods, T., Bos, E., Sutton, G., & Blackett, M. (2013). Short Food Supply Chains and Local Food Systems in the EU. A State of Play of their Socio-Economic Characteristics, *No JRC80420, JRC Working Papers,* Joint Research Centre (Seville site).
Kostof, S. (1991). *The city shaped.* London, Thames and Hudson: Urban Patterns and Meanings Through History.
Kostof, S. (1992). *The city assembled: The elements of urban form through history*. London: Thames and Hudson.
Lanjouw, J. O., & Lanjouw, P. (2001). The rural non-farm sector: Issues and evidence from developing countries. *Agricultural Economics, 26*(1), 1–23.
Lapping, M. B., & Furuseth, O. J. (1999). Introduction and overview. In O. J. Furuseth & M. B. Lapping (Eds.), *Contested countryside: The rural urban fringe in North America* (pp. 1–5). Aldershot: Ashgate.
Laquian, A. A. (2005). Beyo*nd metropolis: The planning and governance of Asia's mega-urban regions.* Johns Hopkins University Press.
Lee, K., Scott, N., & Packer, J. (2014). Habitus and food lifestyle: In-destination activity participation of slow food members. *Annals of Tourism Research, 48,* 207–220.
Lengnick, L., Miller, M., & Marten, G. G. (2015). Metropolitan foodsheds: A resilient response to the climate change challenge? *Journal of Environmental Studies and Sciences, 5*(4), 573–592.
Long, L. M. (2012). Culinary tourism. In J. M. Pilcher (Ed.), *The oxford handbook of food history* (pp. 389–406). Oxford: Oxford University Press.
Losada, H., Martínez, H., Vieyra, J., Pealing, R., Zavala, R., & Cortés, J. (1998). Urban agriculture in the metropolitan zone of Mexico City: Changes over time in urban, suburban and peri-urban areas. *Environment and Urbanization, 10*(2), 37–54.
Madsen, M. F., Kristensen, S. B. P., Fertner, C., Busck, A. G., & Jørgensen, G. (2010). Urbanisation of rural areas: A case study from Jutland Denmark. *Geografisk Tidsskrift-Danish Journal of Geography, 110*(1), 47–63.
Marsden, T. (2013). Sustainable place-making for sustainability science: The contested case of agri-food and urban–rural relations. *Sustainability Science, 8*(2), 213–226.

Marsden, T., & Morley, A. S. (Eds.). (2014). *Sustainable food systems: Building a new paradigm*. Earthscan Food and Agriculture, London: Routledge.
Marsh, R. (1998). Building on traditional gardening to improve household food security. *Food Nutrition and Agriculture,* 4–14.
Maye, D., Holloway, L., & Kneafsey, M. (Eds.). (2007). *Alternative food geographies: Representation and practice*. Oxford: Elsevier.
Mayer, H., & Knox, P. L. (2006). Slow cities: Sustainable places in a fast world. *Journal of Urban Affairs, 28*(4), 321–334.
McClintock, N. (2010). Why farm the city? Theorizing urban agriculture through a lens of metabolic rift. *Cambridge Journal of Regions, Economy and Society, 3*(2), 191–207.
McHarg, I. (1969). *Design with nature*. Garden City, NY, Doubleday: Natural History Press.
Meiberger, E., & Weichbold, M. (2010). How can mountain quality food reduce the vulnerability of mountain farming systems. In *9th European IFSA Symposium*. Vienna (Austria), 1626–1635.
Meller, H. (1994). *Patrick Geddes: Social evolutionist and city planner*. London and New York: Routledge.
Miele, M., & Murdoch, J. (2002). The practical aesthetics of traditional cuisines: slow food in Tuscany. *Sociologia Ruralis, 42*(4), 312–328.
Mok, H., Williamson, V. G., Grove, J. R., Burry, K. S., Barker, F., & Hamilton, A. J. (2014). Strawberry fields forever? Urban agriculture in developed countries: A review. *Agronomy for Sustainable Development, 34*(1), 21–43.
Morgan, K. (2009). Feeding the city: The challenge of urban food planning. *International Planning Studies, 14*(4), 341–348.
Mougeot, L. J. A. (1994). African city farming from a world perspective. In A. G. Egziabher, D. Lee-Smith, D. G. Maxwell, P. A. Memon, L. J. A. Mougeot, & C. J. Sawio (Eds.), *Cities feeding people: An examination of urban agriculture in East Africa* (pp. 1–24). Ottawa: International Development Research Centre.
Munoz, F. (2003). Lock living: Urban sprawl in Mediterranean cities. *Cities, 20*(6), 381–385.
Murdoch, J., & Marsden, T. (2013). *Reconstituting rurality. class, community and power in the development process.* London. New York: Routledge.
Neill, W. J. V., & Schlappa, H. (Eds.). (2016). *Future directions for the european shrinking city*. New York: Routledge (RTPI Library Series).
Neuman, M. (2000). Regional design: Recovering a great landscape architecture and urban planning tradition. *Landscape and Urban Planning, 47*(3–4), 115–128.
O'Rourke, E., Charbonneau, M., & Poinsot, Y. (2016). High nature value mountain farming systems in Europe: Case studies from the Atlantic Pyrenees, France and the Kerry Uplands, Ireland. *Journal of Rural Studies, 46,* 47–59.
Opitz, I., Berges, R., Piorr, A., & Krikser, T. (2016). Contributing to food security in urban areas: differences between urban agriculture and peri-urban agriculture in the Global North. *Agriculture and Human Values, 33*(2), 341–358.
Orsini, S. (2011). Deregulating 'the rural' threatening land management regime. Experiences of space in the Tuscan countryside: arXiv preprint arXiv:1106.0803.
Orsini, S. (2013). Landscape polarisation, hobby farmers and a valuable hill in Tuscany: understanding landscape dynamics in a peri-urban context. *Geografisk Tidsskrift-Danish Journal of Geography, 113*(1), 1–12.
Parasecoli, F., & de Abreu e Lima, P. (2012). Eat your way through culture: Gastronomic tourism as performance and bodily experience. In S. Fullagar, K. Markwell & E. Wilson (Eds.), *Slow tourism: Experiences and mobilities*. Bristol, Buffalo, Toronto: Channel View Publications.
Parham, S. (1992). Gastronomic strategies for Australian citiès. *Urban Futures* 2(2) Canberra.
Parham, S. (1993a). Gastronomy and urban form: South Australian winter planning seminar. In Planning education foundation papers, Adelaide.
Parham, S. (1993b). Convivial green space. In: Proceedings, Canberra: Seventh Australian Symposium of Gastronomy.
Parham, S. (1995). Megalopolis, *Arena* 16 April/May Melbourne.

Parham, S. (1996). *Productive land-use on the urban fringe: A comparative study in planning for regional economic development in Languedoc and Tuscany* (Report). South Australia: Department of Housing and Urban Development.
Parham, S. (2012). *Market place: Food quarters*. Design and Urban Renewal in London: Newcastle upon Tyne, Cambridge Scholars Publishing.
Parham, S. (2014). *Urbanism in a rural context: Is conviviality the missing ingredient? Reflection essay for symposium: how can "new ruralism" and "rural urbanism" contribute to the revitalisation of our countrysides and food systems?*. Sweden: Engelsbergs.
Parham, S. (2015). *Food and urbanism*. London: Bloomsbury.
Parham, S. (2016). Shrinking cities and food. In H. Schlappa & W. Neill (Eds.), *Future directions for the European shrinking city* (pp. 95–113). New York: Routledge.
Parham, S., & Abelman, J. (2018). Food, landscape, and urban design. In T. Waterman (Ed.), *Zeunert, J.* Routledge Handbook of Landscape and Food: Routledge.
Parham, S., & McCabe, B. (2016). *Making space for food in hatfield* (p. 70). Great Britain, University of Hertfordshire Press.
Parrott, N., Wilson, N., & Murdoch, J. (2002). Spatializing quality: Regional protection and the alternative geography of food. *European Urban and Regional Studies, 9*(3), 241–261.
Paül, V., & McKenzie, F. H. (2013). Peri-urban farmland conservation and development of alternative food networks: Insights from a case-study area in metropolitan Barcelona (Catalonia, Spain). *Land Use Policy, 30*(1), 94–105.
Paül, V., & Tonts, M. (2005). Containing urban sprawl: Trends in land use and spatial planning in the metropolitan region of Barcelona. *Journal of Environmental Planning and Management, 48*(1), 7–35.
Paxson, H. (2005). Slow food in a fat society: Satisfying ethical appetites. *Gastronomica, 5*(1), 14–18.
Paxson, H. (2010). Locating value in Artisan cheese: Reverse engineering terroir for new world landscapes. *American Anthropologist, 112*(3), 444–445.
Peters, C. J., Bills, N. L., Wilkins, J. L., & Fick, G. W. (2009). Foodshed analysis and its relevance to sustainability. *Renewable Agriculture and Food Systems, 24*(01), 1–7.
Petrini, C., Furlan, C., & Hunt, J. (2005). *Slow food nation: Why our food should be good. clean, and fair.* New York: Rizzoli.
Phelps, N. A., Parsons, N., Ballas, D., & Dowling, N. (2006). *Post-suburban Europe: planning and politics at the margins of Europe's capital cities.* Palgrave Macmillan.
Pothukuchi, K. (2009). Community and regional food planning: Building institutional support in the United States. *International Planning Studies, 14*(4), 349–367.
Pothukuchi, K., & Kaufman, J. L. (2000). The food system. *Journal of the American Planning Association, 66*(2), 113–124.
Præstholm, S., & Kristensen, S. P. (2007). Farmers as initiators and farms as attractors for non-agricultural economic activities in peri-urban areas in Denmark. *Geografisk Tidsskrift-Danish Journal of Geography, 107*(2), 13–27.
Qviström, M. (2007). Landscapes out of order: studying the inner urban fringe beyond the rural–urban divide. *Geografiska Annaler: Series B, Human Geography, 89*(3), 269–282.
Reardon, T., Berdegué, J., Barrett, C. B., & Stamoulis, K. (2007). *Household income diversification into rural nonfarm activities*. Transforming the Rural Non-farm Economy: Baltimore MD, Johns Hopkins University Press.
Redwood, M. (Ed.). (2009). *Agriculture in urban planning; generating livelihoods and food security*. Abingdon: Routledge.
Rees, W. (1992). Ecological footprints and appropriate carrying capacity: What urban economics leaves out. *Environment and Urbanization, 4*(2), 121–130.
Regoli, F., Vittuari, M., & Segrè, A. (2011). Policy options for sustainability, a preliminary appraisal of rural tourism in Romania: The case of Maramureş. *Food, Agri-Culture and Tourism* (pp. 41–55). Berlin, Heidelberg: Springer.

Renting, H., Marsden, T. K., & Banks, J. (2003). Understanding alternative food networks: exploring the role of short food supply chains in rural development. *Environment and Planning A, 35*(3), 393–411.

Richards, G. (2015). Evolving gastronomic experiences: From food to foodies to foodscapes. *Journal of Gastronomy and Tourism, 1*(1), 5–17.

Richardson, H., & Chang-Hee, C. B. (2017). *Urban sprawl in Western Europe and the United State*. New York: Routledge.

Roden, C. (1989). *The food of Italy*. London. Melbourne: Arrow Books.

Roe, M., Sarlöv Herlin, I., & Speak, S. (2016). Identity, food and landscape character in the urban context: Food and landscape. *Landscape Research*, *41*(7), 757–772.

Roux, A., & Roux, M. (1989). *The Roux Brothers: French country cooking*. London: Basingstoke, Papaermac.

Salvati, L., & Sabbi, A. (2011). Exploring long-term land cover changes in an urban region of southern Europe. *International Journal of Sustainable Development and World Ecology, 18*(4), 273–282.

Scarpato, R. (2002). Sustainable gastronomy as a tourist product. In A. M. Hjalager & G. Richards (Eds.), *Tourism and gastronomy* (pp. 132–152). Psychology Press.

Schmidt, S., Nemeth, J., & Botsford, E. (2011). The evolution of privately owned public spaces in New York City. *Urban Design International, 16*(4), 270–284.

Scott, D., & Duncan, T. (2016). Back to the future: The affective power of food in reconstructing a tourist imaginary. In I. Yeoman, B. U. McMahon, K. Fields, J. Albrecht & K. Meethan (Eds.), *The future of food tourism: foodies, experiences, exclusivity, visions and political capital.* Bristol, Buffalo, Toronto: Channel View Publications.

Selman, P. (2002). Multi-function landscape plans: A missing link in sustainability planning? *Local Environment, 7*(3), 283–294.

Shucksmith, M., & Rønningen, K. (2011). The uplands after neoliberalism?—The role of the small farm in rural sustainability. *Journal of Rural Studies., 27*(3), 275–287.

Sidali, K. L., Spiller, A., & Schulze, B. (Eds.). (2011). *Food, agriculture and tourism: Linking local gastronomy and rural tourism: Interdisciplinary perspectives*. Berlin Heidelberg: Springer.

Simon, D., McGregor, D. & Thompson, D. (2006). Contemporary perspectives on the peri-urban zones of cities in developing areas. In F. M. Duncan & D. McGregor, (Eds.), *The Peri-urban interface: Approaches to sustainable natural and human resource use* (pp. 3–17). London: Earthscan.

Sims, R. (2009). Food, place and authenticity: local food and the sustainable tourism experience. *Journal of Sustainable Tourism, 17*(3), 321–336.

Sonnino, R., & Marsden, T. (2006). Beyond the divide: Rethinking relationships between alternative and conventional food networks in Europe. *Journal of Economic Geography, 6*(2), 181–199.

Stilgoe, J. (1988). *Borderlands*. Origins of the American suburb: Yale University.

Tachieva, G. (2011). *Sprawl repair manual*. Washington DC: Island Press.

Tacoli, C. (1998). Rural-urban interactions: A guide to the literature. *Environment and Urbanization, 10*(1), 147–166.

Talen, E. (2002). Help for urban planning: The transect strategy. *Journal of Urban Design, 7*(3), 293–312.

Tedesco, C., Petita, C., Billen, G., Garnier, J., & Personne, E. (2017). Potential for recoupling production and consumption in peri-urban territories: The case-study of the Saclay plateau near Paris, France. *Food Policy, 69,* 35–45.

Tegegne, A., Mekasha, Y., Tadesse, M., & Yami, A. (2006). Market-oriented urban and Peri-Urban dairy systems. In R. van Veenhuizen (Ed.), *Cities: Farming for the future*. The Philippines: International Institute of Rural Reconstruction and ETC Urban Agriculture.

Tjallingii, S. P. (2000). Ecology on the edge: Landscape and ecology between town and country. *Landscape and Urban Planning, 48*(3), 103–119.

Torres, H., Alves, H., & De Oliveira, M. A. (2007). São Paulo peri-urban dynamics: Some social causes and environmental consequences. *Environment and Urbanization, 19*(1), 207–223.
Trefon, T. (2009). Hinges and Fringes: Conceptualising the Peri-Urban. In F. Locatelli & P. Nugent (eds.). *Central Africa in African Cities: Competing claims on urban spaces,* Leiden Konicklijke: Brill.
Urry, J., & Larsen, J. (2011). *The tourist gaze 3.0.* Sage Publications Limited.
Vall-Casas, P., Koschinsky, J., & Mendoza, C. (2011). Retrofitting suburbia through pre-urban patterns: Introducing a European perspective. *Urban Design International, 16*(3), 171–187.
van Otterloo, A. H. (2005). Fast food and slow food. The fastening food chain and recurrent countertrends in Europe and the Netherlands (1890–1990). In *Land, shops and kitchens: Technology and the food chain in twentieth century Europe* (pp. 255–278). Turnhout: Brepols.
Verhoeve, A., Dewaelheyns, V., Kerselaers, E., Rogge, E., & Gulinck, H. (2015). Virtual farmland: Grasping the occupation of agricultural land by non-agricultural land uses. *Land Use Policy, 42,* 547–556.
Viljoen, A., & Wiskerke, J. S. C. (Eds.). (2012). *Sustainable food planning: Evolving theory and practice*. The Netherlands: Wageningen Academic Publishers.
Vitiello, D., & Brinkley, C. (2014). The Hidden History of Food System Planning. *The Journal of Planning History, 13*(2).
Wackernagel, M., & Rees, W. (1996). *Our ecological footprint: Reducing human impact on the earth*. Gabriola: New Society Publishers.
Wheeler, S., & Beatley, T. (Eds.). (2014). *The sustainable urban development reader* (3rd ed.). London and New York: Routledge Urban Reader Series.
Wiskerke, J. (2009). On places lost and places regained: Reflections on the alternative food geography and sustainable regional development. *International Planning Studies, 14*(4).
Wiskerke, J., & Viljoen, A. (2012). Sustainable urban food provisioning: Challenges for scientists, policy makers, planners and designers. In A. Viljoen & J. S. C. Wiskerke (Eds.), *Sustainable food planning: Evolving theory and practice*. The Netherlands: Wageningen Academic Publishers.
Yokohari, M., Amati, M., Bolthouse, J., & Kurita, H. (2010). Restoring agricultural landscapes in shrinking cities: Re-inventing traditional concepts in Japanese planning. In J. Primdahl & S. Swaffield (Eds.), *Globalisation and agricultural landscapes change patterns and policy trends in developed countries* (pp. 225–244). Cambridge, UK: Cambridge University Press.
Zasada, I. (2011). Multifunctional peri-urban agriculture: A review of societal demands and the provision of goods and services by farming. *Land Use Policy, 28*(4), 639–648.
Zasada, I., Fertner, C., Piorr, A. & Nielsen, T. S. (2011). Peri-urbanisation and multifunctional adaptation of agriculture around Copenhagen, *Geografisk Tidsskrift-Danish Journal of Geography, 111*(1) (Food beyond the city).
Zasada, I., Schmutz, U., Wascher, D., Kneafsey, M., Corsi, S., Mazzocchi, C., et al. (2017). Analysing foodsheds and self-sufficiency for different food system scenarios in European metropolitan regions. *City, Culture and Society*. https://doi.org/10.1016/j.ccs.2017.06.002.

Susan Parham is Head of Urbanism and Planning at the University of Hertfordshire and Academic Director of the International Garden Cities Institute (IGCI). She researches and teaches on placemaking including food and urban design, planned settlements, sustainable materials, and masterplanning and retrofitting. Susan's most recent book is Food and Urbanism (Bloomsbury, 2015) and her latest book chapters are in The Routledge Handbook of Landscape and Food (2018), and in Future Directions for the European Shrinking City (RTPI Library Series, 2016). Susan is a member of the Royal Society for the Arts and the Royal Town Planning Institute.

Part III
Agrourbanism Forms and Places

Chapter 9
Urban Agriculture Forms in Europe

Frank Lohrberg

9.1 Introduction

From 2012 to 2016 the European Commission funded a "COST Action Urban Agriculture Europe". COST stands for "European Cooperation in Science and Technology". It is a pan-European intergovernmental framework strengthening Europe's research and innovation capacities by allowing researchers, scholars, engineers, and practitioners to jointly develop their ideas and take new initiatives. It is not a format for basic research but produces knowledge by meetings, excursions, and periodic exchange of experts in working groups.

The aim of the COST Action was to gain deeper understanding of Europe's different forms of *Urban Agriculture* and to develop a common language, in order to better identify and communicate the potentials of *Urban Agriculture* from a European perspective (Lohrberg et al. 2016). The Action structured its work along an "expedition" to seven places exemplary of Europe's Urban Agriculture. Combined with meetings and conferences, the Action visited the urban regions of Barcelona, Dublin, Geneva, Milan, the Ruhr Metropolis, Sofia, and Warsaw. In addition, the Action examined these and other regions through *Short-Term Scientific Missions* focussing on specific thematic issues. Thus, a first vocabulary of *Urban Agriculture* was found and specified by more than 200 case studies the COST Action collected and made accessible online in the "Atlas of Urban Agriculture[1]".

In this chapter the author shows the definition and typology of urban agriculture as elaborated by the COST Action. A distinction is made between urban farming

[1]See: http://www.urban-agriculture-europe.org/online-atlas.html. Accessed 27 September 2017.

F. Lohrberg (✉)
Department of Landscape Architecture Aachen, RWTH Aachen University, Aachen, Germany
e-mail: lohrberg@la.rwth-aachen.de

E. Gottero (ed.), *Agrourbanism*, GeoJournal Library 124,
https://doi.org/10.1007/978-3-319-95576-6_9

and urban food gardening. A special emphasis is given to map the distribution of these different forms all over Europe, its traditions and aspects of innovation. Doing so, policy recommendation are given with regard to regional differences especially between Eastern and Western Europe.

9.2 Definition and Typology of Urban Agriculture

Based on these activities and existing research, the COST Action defined *Urban Agriculture* as spanning «*all actors, communities, activities, places, and economies that focus on biological production in a spatial context, which-according to local standards-is categorized as 'urban'*» (Vejre et al. 2016, p. 21).

This definition reflects the wide range of urbanisation forms in Europe, which is not characterized by a simple contrast between cities and their rural surroundings but multiple forms of overlays and polycentric structures. For the same reason, *Urban Agriculture* was moreover defined as taking place "in intra- and periurban areas" (*ibidem*), which allows to look at both, the cities' centres and their fringes. Beside the spatial dimension the COST Action emphasized a structural feature of Urban Agriculture as a defining criteria: «*One of its key characteristics is that it is more deeply integrated in the urban system compared to other agriculture. Urban Agriculture is structurally embedded in the urban fabric; it is integrated into the social and cultural life, the economics, and the metabolism of the city*» (*ibidem*).

Expressed the other way round, *Urban Agriculture* is not a rural leftover. As it is interacting with the city, its people, markets and flows. *Urban Agriculture* is part of the city and its growth—not its contradiction. Hence, finding the best forms of interaction paves the way to unfold the high potentials of *Urban Agriculture* in different arenas.

Based on these definitions the COST Action identified fifteen types of Urban Agriculture sorted in three categories (see Fig. 9.1). (1) "Urban food gardening" addresses mainly intraurban areas, where small plots are used individually as allotments and family gardens or collective by community gardens or other forms as therapeutic or educational gardens. (2) "Urban farming" is concentrated in the peri-urban areas and can be distinguished by its purposes ranging from local food production, leisure and education to heritage and environmental aims. (3) "Non-urban farming" characterises an agriculture with marginal interactions to the city, linked only by its spatial structure. These farms have not adapted (yet) to the cities markets and flows but address to global markets. However, due to their location at the urban fringe they have to be considered when talking about future city planning and further *Urban Agriculture* policies.

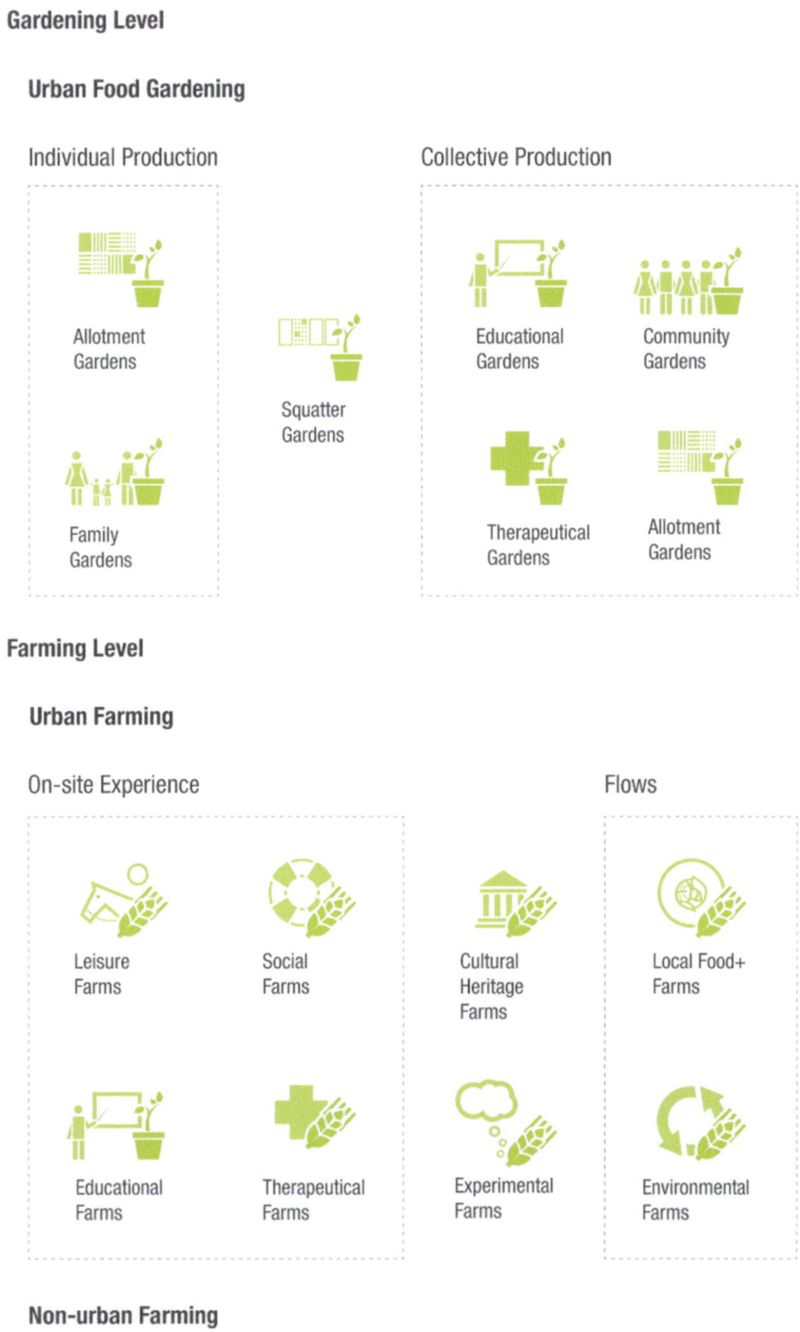

Fig. 9.1 Typology of urban agriculture Europe (*Source* Simon-Rojo et al. 2016, p. 23)

9.3 Distribution of Forms

Agriculture is one of the European Union's central issues receiving almost 40% of its budgets. In particular, the *Common Agricultural Policies* (CAP) with a yearly budget of about 50 billion € heavily influences farmers' decisions all over the continent. As the CAP intertwines agriculture with rural development policies Urban Agriculture is not one of its central aims, but is addressed indirectly when promoting smaller (family) farms and a more specialized, cooperative and sustainable way of production (Lohrberg and Callau 2016). To better support a desirable engagement of the CAP for Urban Agriculture and to figure out which innovations Urban Agriculture could provide for the whole sector, the COST Action aimed at sketching a first picture of Europe's Urban Agriculture.

Following its general typology the Cost Action came up with two maps on different levels—the gardening and the urban farming level. Both maps take the same scheme. Issues of innovation are plotted against a background showing how agriculture has developed in the long-run. The maps roughly identify commons and differences in the distribution of forms throughout the continent. What are the large-scale traditions, settings, and innovation trends that policies should start from?

9.3.1 Mapping Urban Food Gardening

Food gardening has always shaped Europe's cities. Two main traditions can be identified: an allotment tradition based on welfare approaches of the first and second sector and following a collective scheme (1), a food gardening tradition characterized by individual citizens making use of their own garden, or a rented or just-claimed piece of land (2). As Fig. 9.2 shows, some countries are mainly characterized by the first, others by the second tradition and some can be defined as in between.

About the first one scheme, Europe was the centre of the Industrial Revolution. When the industrialisation heavily impacted the cities, gardening activities changed as well. Allotment gardens became popular, a European feature compared with other global regions (Bryant et al. 2016).

As the factories and steelworks attracted a great number of low-paid workers, food security became a crucial issue for industrialized cities. Poor labourers looked for land to grow food and many cities reacted by offering workers' settlements that incorporated food gardens. Howard's (1902) idea of a garden city, which combined suburban housing with different agricultural activities, influenced urbanization processes across the continent.

In the very beginning, allotments were damned as illegal land occupation, but later —especially after the First World War—became established as an official land use form within European cities. The German example clearly shows the strong relation between the rise of food gardening and wars and economic crisis (Lohrberg 2011).

Fig. 9.2 Mapping of European traditions and innovation in urban food gardening (*Source* Lohrberg and Simon-Rojo 2016: 199)

Hence, the distribution of allotment gardens followed mainly the heavy industrialization of the late 19th century. In some countries like Poland, the Czech Republic, Germany or the UK, this urban food garden type spread to other cities and thus became a common phenomenon.

However, in countries like Ireland, Spain, Italy or Greece, the allotment gardens remained a subordinate phenomenon. This may be because the new dwellers still could revert to family-based rural-urban networks. Extended family played a primary role of social support, and now and then, those relatives remaining in villages provided food directly to urbanites.

In a similar way, in these and other countries, family gardens have played a decisive role for urban food production. Up to today, Bulgarian, Slovenian and Slovakian cities are an example of a widespread tradition in using urban and suburban gardens for producing food. As most Eastern European countries reshaped their agricultural system radically, leading to the decline of small, privately owned urban farms, the family gardens gained importance not only for self-consumption but also to partly serve the local markets. In Sofia, it was the vital culture of family gardening helped the citizens to overcome two severe food crises in the early 1990s.

9.3.2 Aspects of Innovation in Urban Food Gardening

Starting from these two traditions in urban food gardening, the COST Action detected some new developments and innovations. Astonishingly, while urban farming innovation (see below) mainly takes place in Western regions of Europe, urban food gardening gets new impulses from all over the continent. They all result from a general interest of city dwellers in reintegrating nature, greenspace and food production as part of their daily life. This notion is part of a general change towards more sustainable urban lifestyles, but the financial crisis of 2008 greatly accelerated the trend. It led to a loss of income and a decreasing standard of life, especially in the South and East of Europe, and raised the question of sustainable development all over the continent.

In this regard, most conspicuous is the rise of "community gardens" as a new type of urban food gardening. According to Simon-Rojo et al. (2016) community gardens are characterized by their collective organization. «*They are not only about growing vegetables, but also about growing social networks, building meetings places, and establishing a sense of community*» (Simon-Rojo et al. 2016, p. 25). Initiatives started in the metropolises (e.g. Paris, London, Berlin, and Vienna) but spread out all over Europe. In countries with an allotment garden tradition, the community gardens often define themselves in contrast to the traditional type: while allotments are considered as an unpolitical retreat, the community gardeners see themselves part of a transition process.

In this process, the innovation is not so much in food production, but in the governance of the city. Community gardens clearly show the potentials of an active citizenship. They support community building and improve neighbourhoods, something the administration is often increasingly struggling with. In Malmö (Sweden), for example, in the "*Odlingsnätverket Seved Ideell Förening*" project a real estate company initiates community gardens to make deprived areas more attractive with targeted side effects for decreased criminality and vandalism and increased real estate values (van der Schans et al. 2016). Meanwhile in Spain, community gardens are a major part of local initiatives as nodes of ecological literacy, to reclaim public space, and to set up new strategies for sustainable urban food systems and food sovereignty (Simon-Rojo et al. 2016).

However, searching innovation only in community garden initiatives risks overseeing a second food gardening trend, which is the renaissance of allotment gardens. Astonishingly, this "comeback" also takes place in countries with no strong allotment tradition. In Ireland or Portugal, municipalities engage in new allotment gardens. The gardens are on a legal base, supported by the municipalities and facilitated with water and so on. Thus, they can be addressed as traditional allotment gardens, but they show notable innovations. In Cascais (Portugal) the new allotment gardens are partially embedded in broader recreation areas, and the municipality aims to shift the food production from subsistence-oriented to more professional business models. Similar goals are addressed by a high an increasing number of new municipal allotment initiatives reported from the Attica region in Greece. In Tartu, Estonia, the municipality created an initiative for low-cost allotments on abandoned development sites aiming at both offering cheap access to a food garden for the broad public, and improving the neighbourhoods. To the same end, in Rabot-Blaisantvest (Belgium) the loan for allotment gardens can be paid by using a local social coin, called the "Toreke". It can be earned through volunteer-work, thereby stimulating both local and sustainable consumption and community building.

European policies should be aware of both the rise of community gardens and the reinvention of allotment gardens. As the case of Spain shows, these food gardens are not only once-off incidents, but shape the cities significantly; between 2006 and 2014, the amount of plots in allotment and community gardens has increased by six (Ballesteros 2014). They clearly demonstrate a shift in urban society, as people use food to regain control of their daily lives, either in terms of food security or in terms of engaging in more sustainable ways of urban living. In light of the economic and social challenges food gardening is revealing, its policy potentials should be more clearly addressed.

E.g., urban food gardening is seldom set up as a professional business from the beginning. As the plots needed are small and—compared to urban farming—investments are low, it can occur spontaneously and in the short-term. That is why the economic crises from 2008 pushed especially these types of Urban Agriculture. In many cases, there is evidence that it is just uncertainty that stimulates new initiatives. As a low threshold undertaking, the food gardens hold a notable potential for offering new labour, income and education. Once established, it may only be a small step from non-profit food gardening to new businesses, as shown by *the Rotterdam Munse project in the Netherlands.*[2] There, women from diverse ethnic and social origins grow herbs and sell this to local restaurants and cafés. As shown here, urban food gardens can be a breeding ground for sustainable economic development. To benefit from this, further investigation, communication and networking of Europe's food gardening experience is needed. Instead of mitigating social problems by classical welfare measures (e.g. food banks), urban food gardening provides a way to empower the people themselves—a major precondition for managing economic crisis and social challenges effectively.

9.3.3 Mapping the Evolution of Urban Farming

Returning to the influence of industrialization, there is also a European feature at the urban farming level. Back in the eighteenth and nineteenth centuries, when industrialization moved from the rural mining areas to the growing cities, the situation for farming changed substantially. Up until then, farms in and near cities delivered all kinds of food and produce needed by the citizens. However, due to railways and steamships, international food systems developed, and urban farms lost much of their advantage in producing close to the urban market. How did farms react? One opportunity was to go the 'rural way', trying to enlarge farms and produce cash crops for the global market—the "non-urban farming" category as pointed out above. The number of farms involved here is decreasing (as farms have to grow in order to compete), but they use the land intensely and without abandonment. The other option was to specialize in niches that could still benefit from proximity to the city, principally horticulture and market gardening (Spitzer 1974)—"the urban farming" category. Farmers grew fruits, vegetables, and flowers on the small plots that urbanization did not take up and achieved high prices when direct selling at the nearby urban markets.

Since the 1970s, environmental issues have gained momentum, leading to a more significant development of the share of organic farming. Subsequently, over the course of the following decades, urban farming reacted to emerging demands and trends, such as leisure (e.g., equestrian sport), health concerns, or biodiversity.

Where can we find an urban farming tradition like this in Europe? Figure 9.3 shows that the adaptation process has been quite different throughout the continent in the last decades. We can find countries with a specific history in urban farming especially in the Western and Central parts of the continent, stretching from the Benelux to Italy.

Why this? As urban farming mostly needs many cost intensive facilities (machines, buildings, irrigation, etc.) most business models are only successful in the long run. To this end, the farmers need to reliably predict future economic and

[2]See: www.rotterdamse-munt.nl. Accessed 14/02/2018.

Fig. 9.3 Mapping of European traditions and innovation in urban farming (*Source* Lohrberg and Simon-Rojo 2016, p. 195)

spatial development in order to make the right decisions. For example, no farmer will invest in a new facility if the area could soon be turned into a housing development. However, if processes are predictable, business models differentiate and specialization happens, leading to a wide range of urban farming types as described in Fig. 9.1. In this process, the growth of the city does not automatically lead to the end of agriculture. In contrast, new fringes—caused by urbanisation—lead to new contact zones offering opportunities for farmers who are capable of adapting.

Back to the question of divergent distribution in Europe, we can assume that an ongoing but somehow controlled urbanization benefits urban farming. Especially in Western European countries' national, regional or local bodies decided to channel the urbanization from the outskirts to an intra-urban redevelopment. For example, in the Canton of Zurich, the people voted in 2012 for a "kulturlandinitiative" prohibiting any further claiming of the Canton's 40,000 ha of agricultural land. To operationalize such ambitious targets, elaborated planning systems are at hand and able to regulate the land use. Zoning concepts sometimes hinder agriculture's expansion but on the other hand, they safeguard agriculture in distinct areas, thus allowing long-term investments there. The firm protection prevents the rise in land prices associated with the expected return of potential "urban developments".

Ultimately, urbanisation is still challenging the farmers but with the processes as described above, they are able to benefit from the opportunities the urban realm is creating. Thus, over the last decades in these countries, urban farming has reacted to the diverse demands of the city, starting with food supply but over time encompassing recreational, environmental, social and other purposes as well. Thereby, urban farming is often a forerunner in the multifunctionality of farming overall (Bryant and Johnston 1997).

In contrast, we cannot find such a strong process of ongoing adaptation outside Europe's central regions, especially not in Eastern Europe. We can assume that there must have been similar traditions of serving the cities' markets through surrounding agriculture, however, socialist economic systems and the restriction of private investment capped or marginalized these traditions in urban farming. Hence, the number of small-scale entrepreneurship being able to benefit from the diverse opportunities of the urban market was low. Since the end of cold war, those opportunities for farmers emerged but could not be turned into successful businesses due to a high grade of uncertainty. As the Warsaw or Sofia case show, the planning system shifted from regulation to deregulation and liberalisation, causing an unpredictable spatial development. In addition, the property situation was often not clear and the general economic development was quite erratic thereby paralysing urban farming. Thus, in general, innovation in this sector was low. Agriculture gives distinction to most Eastern Europe cities' fringes, but farming is characterized as non-urban adapted.

9.3.4 Aspects of Innovation in Urban Farming

With regard to current innovation, this distinction between East and West is still relevant. We mainly find Urban Agriculture innovation, defined as successful adaptation to urban challenges, in Europe's Western parts. The COST Action elaborated on 4 different paths of innovation:

(1) *new business models*: as explained above, adapting to an urbanizing environment is essential for urban farming. Hence, innovation is an intrinsic characteristic and an ongoing issue for urban farmers. Today, the COST Action identified several innovative "business models" (van der Schans et al. 2016) going beyond what has generally already been achieved, namely direct selling or farming for recreation or biodiversity purposes. For example, in the Geneva region, farmers operate more diversely by interfering with the urban metabolism. They collect waste and turn it to compost or bioenergy (Sage and Tornaghi 2016). Turning the farm into a "social" one is another innovative business model as cases from the Ruhr area (Germany), Rome, the Attica region (Greece) or Cascais (Portugal) demonstrate. According to Simon-Rojo et al. (2016) social farms aim at promoting disadvantaged people's rehabilitation and the integration of people under risk of exclusion by offering them low-threshold jobs in agriculture. «*The target group covers convicts, drug addicts, juvenile offenders, precarious immigrants, and people with 'low contractual capacity'*» (Simon-Rojo et al. 2016, p. 26). Going beyond direct selling and engaging the costumers as co-producers is another promising business model. Farmers leave parts of the production to their customers, e.g. maintaining and harvesting the fruits as the "Meine Ernte",[3] business model from Germany shows More than this self-harvesting by customers, farmers can jointly organize the whole production process together with the customers—as the *Community Supported Agriculture* (CSA) model shows. As this innovation is based on advanced cooperation, it is described under (4).

(2) *territorial based approaches*: some regions have developed a territorial approach to upscale urban farming from a normal business to a city affair. Barcelona (*Baix Llobregat*) has been forerunner in this "agricultural park" idea, aiming first at land protection for an ongoing agriculture and then promoting agriculture itself by economic support, mainly marketing and cooperative direct selling. Similar "park"-approaches now occur in other regions as well (Timpe 2017): In Cologne, the "Belvedere Park" encompasses 300 ha of agricultural land next to the cities green belt, which have been reshaped for city dwellers by new paths and lookout towers, highlighting the farmland as a desirable, vital urban space. In the *Park de Lilas* near Paris different forms of urban agriculture (allotments, community gardens, horticultural production) are framed by a park design out of paths and tree rows to benefit recreation and public access. No matter of different purposes, the agricultural parks contribute significantly to the cities' Green Infrastructure (Timpe et al. 2016) and a sustainable city development in general.

(3) *heritage based approaches*: similar to the territorial approach some projects highlight urban agriculture by its historic features. Often Urban Agriculture is

[3]See: http://www.meine-ernte.de/. Accessed 27 September 2017.

regarded as a new post-millennium phenomenon, but with a closer look in many cities forms of agriculture can be detected that deeply root back for hundreds of years. Valorising this feature as heritage offers an innovative way to manage the agricultural land and benefiting the city at the same time, as the Milan "Parco Sud" (Italy) shows (Branduini et al. 2016). There, the park preserves historic landscape elements (irrigated meadows ("marcita"), rice fields, farmhouses) as the remains of a complex ancient agricultural system dating back to the 13th century. With thematic paths, information boards and a museum the park raises awareness for the historic dimensions of the agrarian landscape. In Bamberg (Germany) the medieval city is protected as World Heritage. One part of the protected zone is the so called "Gärtnerstadt" which means gardeners city. There, since hundreds of years market gardeners live and work on the backstreet fields. During the time, they specialized their production to the needs of the cities market, mainly vegetables. The gardeners associations have formed distinctive traditions keeping the knowledge of cultivated species, recipes and so on and contributing to the cities folk culture by markets and processions. These traditions have been acknowledged as Germany's UNESCO intangible heritage in 2016. Both, tangible and intangible heritage are showcased in different ways. The city of Bamberg set up a museum in an old farmhouse authentically demonstrating the way gardeners worked in former times. Part of the museum is a "garden of species" showing the unique species of onions, liquorice and so on the gardeners have cultivated. A thematic path crosses the Gärtnerstadt and lead to an outlook tower giving sight to the farmers' fields. Similar, in Elche (Spain) the UNESCO designated a huge palm grove in the inner city as World Heritage. The palm grove dates back to Moorish times. Today the palm grove is an essential port of Elche's green infrastructure and its identity, too. E.g., palm tree branches are part of the coat of arms.

(4) *networking*: a fourth form of innovation bases on networking, establishing cooperation and supporting "newcomers" in agriculture. In France, the COST Action identified advanced networking between municipalities and chambers of agriculture, the "Terres en Villes" consortium. Within this, some cities support their local agriculture by economic or environmental programs or by procurement initiatives based on regional products, also leading to a higher willingness of farmers for investments. A major innovation by cooperation is associated with initiatives of *Community Supported Agriculture* (CSA), which bring together professional urban farmers and citizens that do not only operate as customers but also as "co-creators" of the farm. We found an accumulation of CSA approaches around the Alps (Milano, Geneva, Zurich, Munich, Vienna) but other regions are following this scheme with new facets, for example in Southern Spain (Sevilla). Beside existing farmers searching for new cooperation, "newcomers" in Urban Agriculture play an increasing role in fostering innovation. In general, there are high barriers for people owning no land to enter the agricultural business. However, it is exactly this urban environment that creates some niches for successful start-ups. As projects from Switzerland

(Urban Farmers, Basel), the Netherlands (*Daktuin Zuidpark* rooftop farm, Amsterdam) and elsewhere show, stimulating factors are—against expectation —access to land by making productive use of under-utilized resources. Initiatives do not compete with traditional farmers but go for abandoned sites, roof-tops ore public land. E.g., in Vienna, the Association *Grünstern Lobauerinnen*[4] act as a "planting community" on municipal land. In Madrid, as a reaction to the economic crisis, a new group of farmers emerged, resembling forms of Urban Agriculture in the Global South: labour-intensive, small scale, slow expansion, even starting as an informal activity. Meanwhile, their number is larger than those registered officially in the agrarian census (Ballesteros 2014). Urban farming in Eastern Europe lags behinds these trends. However, there are promising starting points. For example, in the Czech Republic and Poland networks of educational farms entered the scene. In Slovakia, in the Bratislava and Nitra regions, an urban farming tradition, partly based on cooperation of farmers, is still vital and has gained momentum in the last decade. In Tartu, Estonia, a modern greenhouse set high standards in ecologically friendly production and in integrating horticultural production in an urban environment.

In general, a European policy on urban farming has to take the different speeds of adaptation and innovation into account. While Western countries would benefit more efficiently from further networking, education in business models, and spatial integration in the urban fabric, in Eastern Europe, policies have to raise awareness of the urban farming issue in general. Policies should improve preconditions for adaptation, e.g. by securing environments for agricultural investments through zoning or initiating an urban farming orientated procurement. For the prosperity of urban farming, the predictability of development, both in spatial and economic terms, is crucial.

9.3.5 Conclusions

Within the COST Action Urban Agriculture Europe different types and forms of urban agriculture could be identified. By mapping traditions and innovations a clearer picture of urban agriculture was sketched. On the food gardening level two lines of tradition were detected, allotment gardens and family gardens. Nowadays, municipalities all over Europe rediscover the advantages of allotment gardens. In parallel, community gardens have established as a new type of urban food gardening. Their main achievement is not in food production for its own sake but in making use of food production for other mainly social and environmental purposes.

On the urban farming level adaptation processes turned out to be a decisive feature. The growing of cities offers new markets for farmers, but only if the

[4]See: http: www.gruenstern-lobauerinnen.at. Accessed 27 September 2017.

urbanisation is channelled by zoning concepts the farmers will make us of the upcoming opportunities by long-term invests. Interestingly, while urban food gardening's evolution can be explained as a reaction to a multiple uncertainty due to various socio-economic crisis, urban farming thrives only when future conditions are predictable. Hence, in Europe most urban farming activities concentrate on Western countries with a longer tradition in restrictive land use policies. There, new business models, new territorial, thematic and networking approaches make urban farming a promising tool for local city planning and European policies as well.

References

Ballesteros, G. (2014). Espectacular crecimiento de los huertos urbanos. *El Ecologista, 81*, 54–55.

Branduini, P., Laviscio, R., Scazzosi, L., Supuka, J., & Tóth, A. (2016). Urban agriculture and cultural heritage: A historical and spatial relationship. In F. Lohrberg, L. Licka, L. Scazzosi, & A. Timpe (Eds.), *Urban agriculture Europe* (pp. 138–147). Berlin: Jovis.

Bryant, C., & Johnston, T. (1997). *Agriculture in the city's countryside*. London: Bellhaven.

Bryant, C., Peña Díaz, J., Keraita, B., Lohrberg, F., & Yokohari, M. (2016). Urban agriculture in a global perspective. In F. Lohrberg, L. Licka, L. Scazzosi, & A. Timpe (Eds.), *Urban agriculture Europe* (pp. 30–37). Berlin: Jovis.

Howard, E. (1902). *The garden city of tomorrow*. London: Swan & Sonnenschein.

Lohrberg, F. (2011). Debatten über die Funktion von städtischem Grün und Freiflächen in der Stadt- und Raumplanung seit dem Kaiserreich. In B. Walter & T. Küster (Eds.), *Aus der Hand in den Mund – Selbstversorgung als Praxis und Vision der modernen Gesellschaft. Westfälische Forschungen* (Vol. 61, pp. 261–278), Münster.

Lohrberg, F., & Callau, S. (2016). The CAP reform as a chance for urban agriculture. In F. In Lohrberg, L. Licka, L. Scazzosi, & A. Timpe (Eds.), *Urban agriculture Europe* (pp. 202–207). Berlin: Jovis.

Lohrberg, F., Licka, L., Scazzosi, L., & Timpe, A. (Eds.). (2016). *Urban agriculture Europe*. Berlin: Jovis.

Lohrberg, F., & Simon-Rojo, M. (2016). Mapping urban agriculture on the European scale. In F. Lohrberg, L. Licka, L. Scazzosi, & A. Timpe (Eds.), *Urban agriculture Europe* (pp. 194–200). Berlin: Jovis.

Sage, D. M., & Tornaghi, C. (2016). Urban agriculture practices on the metabolic frontier: Cases from Geneva and Rotterdam. In F. Lohrberg, L. Licka, L. Scazzosi, & A. Timpe (Eds.), *Urban agriculture Europe* (pp. 178–181). Berlin: Jovis.

Simon-Rojo, M., Recasens, X., Callau, S., Duží, B., Eiter, S., Hernández-Jiménez, V., et al. (2016). From urban farming to urban gardening. In F. Lohrberg, L. Licka, L. Scazzosi, & A. Timpe (Eds.), *Urban agriculture Europe* (pp. 22–28). Berlin: Jovis.

Spitzer, H. (1974). Die Ansprüche der modernen Industriegesellschaft an den Raum, dargestellt an Beispielen der Landwirtschaft im Modellgebiet Rhein-Neckar. In *Die Ansprüche der modernen Industriegesellschaft an den Raum* (Vol. 90, pp. 1–54.). Hanover: Akademie für Raumforschung und Landesplanung.

Timpe, A. (2017). *Produktive Parks entwerfen: Geschichte und aktuelle Praxis biologischer Produktion in europäischen Parks* (Dissertation). Rheinisch-Westfälische Technische Hochschule Aachen.

Timpe, A., Cieszewska, A., Supuka, J., & Tóth, A. (2016). Urban agriculture goes green infrastructure. In F. Lohrberg, L. Licka, L. Scazzosi, & A. Timpe (Eds.), *Urban agriculture Europe* (pp. 126–137). Berlin: Jovis.

van der Schans, J. W., Lorleberg, W., Alfranca, O., Alves, E., Andersson, G., Branduini, P., et al. (2016). It is a business! Business models in urban agriculture. In F. Lohrberg, L. Licka, L. Scazzosi, & A. Timpe (Eds.), *Urban agriculture Europe* (pp. 82–91). Berlin: Jovis.
Vejre, H., Eiter, S., Hernández Jiménez, V., Laviscio, R., Lohrberg, F., Loupa-Ramos, I., et al. (2016). Can agriculture be urban? In F. Lohrberg, L. Licka, L. Scazzosi, & A. Timpe (Eds.), *Urban agriculture Europe* (pp. 18–21). Berlin: Jovis.

Frank Lohrberg born in 1964, studied landscape architecture at Hanover University and won the Peter Joseph Lenné Award in 1990. Beginning in 1994, he worked at the University of Stuttgart where he received his doctorate in 2001, which dealt with urban agriculture and city planning. In 2002, he founded the office lohrberg stadtlandschaftsarchitektur, which focuses on the landscape architecture of regional open space. In 2009, he was appointed to the Chair of Landscape Architecture at RWTH Aachen University. From 2012–16 he chaired the COST Action TD 1106 'Urban Agriculture Europe'.

Chapter 10
Agricultural Park in Europe as Tool for Agri-Urban Policies and Design: A Critical Overview

David Fanfani

10.1 Introduction: Framework of Reference

During recent decades, and with an ongoing evolution, urbanization processes—in both western and Global South countries—have taken place in accordance with growth patterns that not only entailed land being taken in a quantitative way, but that also produced dispersed and low density urban settlements. In such contexts, paradoxically, urban and rural domains intertwine in a way that no longer allows the classical distinction between them to be taken for granted. This Intermediate domain "neither rural nor urban" (Allen 2010; OECD 2009; Wandl et al. 2014) is a concept that calls not only for radical and critical rethinking of planning studies categories and analysis models (Brenner and Schmid 2015a, b), but also for innovative approaches in the fields of politics, spatial planning and design tools.

With such a backdrop, open spaces close to urban and impervious areas, which are by their very nature ecological, agricultural and wooded domains, are no longer seen as separate land to accommodate further urban expansion and services. In fact they are conceived as key endowments and "urban/regional patrimonies" that exist to cope with the challenging trends and factors entailed by global trends such as climate change and post oil transition regime. The latter factors in fact, strongly suggest a radical re-embedding or "re-localization" of urban settlements in their surrounding regions (Thayer 2013; Magnaghi 2010, 2014) in order to "fix" the metabolic rift (Bellamy Foster 1999) between these two domains, lost with the "specialization" of modern age and capitalistic settlements (Polany 1944).

Moreover, the concept of the peri-urban domain as featured with particular characteristics, emerges in this context as a qualitatively distinct one, as a milieu crossed by flows, practices and relationships, that allows it to be conceived as a

D. Fanfani (✉)
DIDA—Architecture Department, Florence University, Florence, Italy
e-mail: david.fanfani@unifi.it

E. Gottero (ed.), *Agrourbanism*, GeoJournal Library 124,
https://doi.org/10.1007/978-3-319-95576-6_10

transitional zone that calls for innovative practices and tools of analysis, planning and design (Allen 2003; Gallent et al. 2006).

Starting from this growing awareness of the relevance of manifold and mutual relationships between human settlement and surrounding agro-ecological domains, discourses on town and country relationships gained momentum (Hidding et al. 2000; Van Leuven 2010). This highlighted issues related not only to the role of rural and ecological areas as a "mold" for the ecologically oriented design of settlement patterns (Tijallingii 2000), but also to the necessity to pursue a "new alliance" between the city and countryside. That in such a way that they mutually co-evolve upon self-relied principles of development (Scott Cato 2013), ecologically, and socially fair balanced relationships (Magnaghi and Fanfani 2009).

This "new alliance" encompasses manifold issues, and is based on a mutuality principle. In fact, the opportunity to beneficiate the urban domains with the manifold eco-systemic services delivered by surrounding rural areas (e.g. flood control and storm water retention, carbon capture, groundwater sanitation, drinking water disposal, micro-climate control, biomass for renewable energy sources, etc.) finds support in the protection and permanence of farmland and agricultural activities. The latter, conversely, finds in the urban market and the demands of new citizens, a (finally) viable market that calls for healthy food production, and for connected and productive innovation (Jacobs 1984) or "retro-innovation" (Stuiver 2006).

These demands, and the new relationship between the urban and rural domain, find at various government levels and on behalf of public bodies, manifold initiatives aimed to innovate. Such methods, practices, and design tools overcome the inherit separation between urban and rural domains, regional/urban planning and rural development and the related routine sector oriented approaches. Some study programs particularly highlights the innovative potentialities of these in-between areas (PLUREL 2011; METREX 2011; ESPON-EDORA 2013).

This set of policy and design studies encompasses many topics and issues, and adopts a cross-sectorial approach in keeping with the multi-dimensional challenges that are present. The role of green and blue networks as "territorial infrastructure" in city regions (Shrijnen 2000), or ecological networks to support human settlements are also strongly addressed for instance at the European level, by other public document and research/action programs. In particular, this refers to their role in delivering either a pleasant environment to live in, enhancing biodiversity and supporting basic *Ecosystem Services* (ES), needed for settlements of life, and is consistent with the goal of achieving a healthy and resilient urban environment (URBES/ICLEI 2015; Millennium Ecosystems Assessment 2005; Green4Grey 2016). In this framework, a new vision has also emerged, with importance related to the role and value of permanent peri-urban agriculture, both as an opportunity to support and recover settlement metabolism and landscape amenities, and to trigger newly sustainable local economies at the urban-rural interface (EESC 2004; AA. VV 2010; Terres en Villes 2008; FAO 2015). Along with this has been the creation and support of some associated realities, aimed at promoting innovative management of land and agriculture in urban and peri-urban areas. Finally, the matter of the recovery of peri-urban agriculture, especially in relation to the creation and support of regional and local food systems, and of complementing planning practices, has found wide coverage in planning reflections (Morgan 2009; Morgan and Sonnino 2010; Rururbal 2011), public debate and documents (European Regions Committee 2011), and in policy and design guidelines (APA 2007; Viljoen et al. 2005), that has also helped pioneered local experiences.

10.2 Genesis, Evolution and Basic Features of the Peri-urban Agricultural Park Model

The previously mentioned pervasive nature of urbanization processes, poses the necessity to respond appropriately to them from a no more exclusive urban and sector focused perspective. It constitutes, as shown, the framework that has inspired, since the beginning of the 1990s, some innovative endeavors in the field of policy, guidelines, tools and practices addressed to treat peri-urban areas planning and design, where urban and rural merge producing an original and "intermediate" milieu. The Agricultural Park model in western countries and especially in Europe, stems progressively from this necessity, and could be considered, although not yet in a codified form, as a tool that is strongly addressed, more than with other types of peri-urban open spaces protective tools, to fully develop the mutual relationships between urban and farmland areas. In particular, in order to foster and enhance farming activities, as the weaker of the two poles, the latter taking advantage of the proximity to the urban environment both as viable market places for food production and as a source of services and incentives for innovation (OECD 2013).

10.2.1 Origins and Development of the Concept

The original conception of the AP can be found in the seminal experience of the Milan Province and metropolitan area, in Italy. There, an important bottom-up movement fostered the institutional constitution of the "Parco Sud Milano"—a large, prime farmland park of more than 11,000 ha. It's aim was to hamper the progressive land taking process in the southern sector of the Milan metro area by fostering the traditional agriculture presence in the area. This endeavor, which in 1992 was finally granted with institutional formalization as a territorial park, was fuelled not only on behalf of social movements, but also by strong engagement from active university field-researchers. The first definition of an AP originated from this context, and was conceived as (Ferraresi and Rossi 1993): «*…a territorial structure aimed at primary production, to its protection and valorization, and assuming a secondary goal of leisure and cultural exploitation on behalf of citizens, while being compatible with the main activity. Both goals give place to different environments and structures in the park area. The valorization of the natural* *environment and the ecosystem balance raise as a necessary requirement in order to obtain the first two*» (Ferraresi and Rossi 1993, pp. 45–46) (author translation).

The strong focus on the protection of farmland pursued by the AP, has also been characterized in the Catalan experience. From the establishment of the first *Parque Agrario de Sabadell* (1993) arose the definition of the Nature Protection Council, and accordingly the AP was conceived as: «...*A confined open space which purpose is to ease and to ensure the farmland exploitation continuity, protecting it from its embodiment it the urban process, promoting specific programs that allows for the development of its natural environment and socio-cultural economic potential and to protect the natural patrimony of surrounding areas*» (Montasell 1996, p. 40).

Further definitions followed, especially in the Spanish/Catalan context (Zazo Moratalla and Yacamán Ochoa 2015, pp. 15–17), that stressed other particularities of the AP model, until, 2004, when the *European Economic and Social Committee* issued its opinion on peri-urban Agriculture (EESC 2004). Te latter was a pivotal document in the evolution of the idea of active protection of peri-urban farmland areas, and for the concept of the AP itself. The document started by pointing out the importance of protection of sustainable peri-urban agriculture, along with the services that it can deliver, such as ecosystem and biodiversity protection, food production, enhancement of the urban environment, etc., and in coping with the growing metropolization process and urban sprawl in Europe.

Starting from this position, the EESC opinion outlined its main goals in order to achieve a shift from weak and threatened peri-urban agriculture, to an enhanced and viable multi-functional activity, suitable to provide healthy food and other key services for local urban areas. The document outlined some key goals in order to enhance these aims:

- The key role of protective land use policies at the municipal and supra-municipal level;
- The need to improve citizen awareness about the importance of peri-urban agriculture and the fragility of the areas where it takes place;
- The importance to develop a viable and dynamic peri-urban agricultural system.

To jointly achieve these goals, the document called for the adoption of participative and steady governance of these areas on behalf of public administration. It also recommended that a commitment should be pursued by those involved, including all stakeholders, citizens and by the various elements of civil society. Finally a governance process was suggested by the document, outlining two main tools:

- A formal cooperation network or "participatory and managerial body", as a stable public-private partnership and governance subject;
- Rur-urban inter-municipal projects, aimed at improving coordination of local projects, with cooperation between different administrations and citizens, private parts, in the context of public funding and rural development tools.

Even without the introduction of a specific label, these two main tools embody the first "profile" of an Agricultural Park operating structure. Added to this, the document outlines an integrated, planning regulatory system aimed to define a territorial project of conservation and development of spaces destined to peri-urban agriculture.

Following the EESC opinion document, there were other contributions which help to outline the main and essential "design" feature of an Agricultural park. These features related not only to its institutional architecture and commitments, but also to the requirements needed to trigger the process of park design, creation and for its progressive development. However it is worth underline that the local context of an AP usually represents and embodies only some, rarely all, of the features that this model proposes.

Other reflections, before and after the EESC opinion was issued, trying to point out a single operational definition of the AP model. This particularly refers to the conditions for the creation, establishment and management of an AP. Along these lines, Montasell—drawing on the field experience gained from the creation of the peri-urban AP of *Baix de Llobregat* in the Barcelona metropolitan area—stated that the main requirements which create the possibility for an AP to exist, depend mainly on five conditions: political support from public authorities as a response to public opinion, real collaboration on behalf of public sectors and appointed officers, support and agreement on behalf of farmers and the agriculture sector, and the proposition of a strategic project and vision for the park to be implemented through the action of a management body (see Zazo Moratalla and Yacamán Ochoa 2015, cit., p. 18).

The AP nature appreciation according to Montasell, is strongly related to the idea of a peri-urban farmland area that is actively protected with the recovery of a profitable and urban market oriented agriculture, and fitting with the enhancement of the culture and heritage of rural settlements (Montasell 1996, cit.). Donadieu also started from experience with French urban agriculture. Although he stressed more the multi-functional nature of the AP, especially in playing an environmental/ ecological role and creating services and loisir spaces for the city, he also stated some points that were similar to Montasell in referring to the conditions for an AP creation. Besides the aforementioned requirements, including the will of public administration and the agreement of stakeholders with inter-sectorial management activity, Donadieu's contribution pointed out the pivotal role of spatial planning in defining binding rules for farmland protection, and also to the importance of the activation of land policies to prevent urban developments in rural areas (Donadieu 2005, 2006, pp. 143–148). Afterwards, in a successive research commitment Magnaghi and Fanfani—starting from the EESC document reference points—conceived the AP as a key innovation planning tool, aimed at producing a "counter-mold" for settlements form, in order to jointly achieve at the regional scale, a polycentric and balanced settlement pattern, as well as viable and multi-functional farming activity (Magnaghi and Fanfani 2009, cit.). In this regard, supported also by field research activity, Magnaghi and Fanfani outline two, not mutually exclusive, "archetypes" of AP generating models: the first being a formally recognized entity, which is top-down originated, and has had binding rules of spatial planning applied. The second a voluntary, and at least initially, a more informal network of stakeholders and civic bodies, that share and uphold a vision and a strategy for peri-urban agriculture and farmland protection (Magnaghi and Fanfani 2009, cit. pp. 26–28).

Although outside of the European context, the experience of *Sustainable Agriculture Education* (SAGE), in the San Francisco bay Area deserves mention. There, SAGE has pioneered since 2006, the *Sunol Agricultural Park experience.*[1] This initiative promoted a 3 ha AP as a public/private partnership, and a key tool to protect farmland from urban expansion, as well as to support small farming enterprises, hobby farming and also *Community Supported Agriculture* (CSA). Its goal was to promote a new collaborative and cognitive relationship between peri-urban farmers and citizens, especially fostering food miles schemes, (including the activities of local retail and shops), educational initiatives, services and structure for park appreciation and use, community solidarity and a "celebration" of agricultural community values.

Returning to Europe, it is worth recall the activity of the *Interreg IVc Programm "Periurban Parks".*[2] This program, unfolded from 2010 to 2012, was mainly focused in defining an operational methodology for peri-urban green space and biodiversity protection and management. Although not exclusively, in this framework attention was posed to the role and nature of the AP and peri-urban agriculture as tools to complement, and embody sustainable strategies, actions and pilot projects, for land and ecosystem protection (see Fig. 10.1).

Thorough and pivotal support in defining the nature, methods of creation, management and duties fulfilled by an AP, can be drawn from Zazo Moratalla (Zazo Moratalla and Yacamán Ochoa 2015; Zazo Moratalla 2015a). In particular, in the second of these works, which starts from a "genealogical" and analytical approach to the Italian and Spanish AP examples, the AP is mainly conceived as a tool to convert a weak and badly managed peri-urban agricultural space (*Espacio Agrario Periurbano*) into a dynamic, viable and self-reliant, agricultural, peri-urban system (SAP: *Sistema Agrario Periurbano*). The AP still maintains its multi-functional role in order to preserve and foster heritage and environmental values, in accordance with the development of innovative and sustainable farming activities. These are especially aimed at setting and promoting a local food system, resulting in two main strategies: reapprochement of producers-consumers for the achievement of local food sovereignty, and sustainability and innovation in productive, commercial and service processes.

---

[1]Starting from 2017 SAGE delegated the Sunol Park management activities to *Alameda County Resources Conservation District* (ACRCD) to manage the sustainable farming and ecological benefit elements and the *San Francisco Public Utilities Commission* (SFPUC) to run school and community education programs.

[2]See: http://www.periurbanparks.eu/live/index.php?a=open&id=4c99fc98837e2&ids=4c8ff07964a15&l=en (Accessed 11/12/2017).

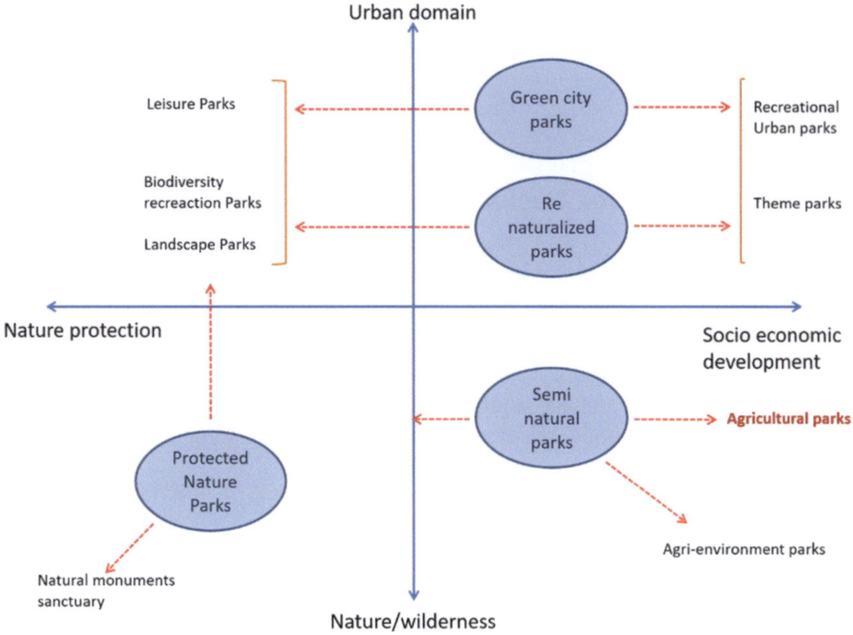

Fig. 10.1 Concept map of peri-urban park typologies according to the 4 typologies identified by the Project (*Source* Fanfani-Peri-urban Parks Project-Interreg IVc 2012)

As well as stating the AP main goals, Zazo Moratalla outlines the key "mechanisms" (*mecanismos*) required to trigger, define, protect and enliven an AP, including with each of them a specific mechanism corresponding to a defined tool (*instrumentos*). The key element in this framework is represented by the governance (*gobernanza*) mechanism, that can be seen to cross—from either a top-down or a bottom-up generated perspective—each element of the AP structure. This occurs either in the initial creation phase or afterwards in the development process. According to the "ideal" model proposed by Zazo Moratalla, the institutionalization of the AP needs to be achieved primarily with a shared vision of the park (*Projecto del Parque Agrario*), along with the provision of a set of tools that relates to the spatial planning strategy of active protection (*Plan de Ordenacion Territorial*) at urban and regional scale, with coordination of the different sectorial plans. Finally, the integrated development strategy needs to be embodied in a special development plan (*Plan Espècial de Desarrollo*) which is responsible for the creation and duties of a combined public-private management body (*Ente Gestor*).

The model of an AP proposed by Moratalla represents a very important result and contours in a systematic way, the very nature and dynamic embodied by an AP, suitable to address the cross-disciplinary matters that arise in managing the planning and design of peri-urban farmland areas. Moreover, the model—without any claims of universal validity—presents a dynamic process structure that allows for suitable application in different contexts, regarding ways to manage operational integration between the different fields of public policies, especially between urban planning and rural development tools.

10.3 Peri-urban Agricultural Parks in the European Context. Some Regional Cases

Growing awareness resulting from disciplinary reflection about the very nature of the peri-urban agricultural park model, is deeply intertwined with real practices and initiatives inspired to this model. The following paragraphs illustrate in a summarized form, and according to a national sampling some of the more interesting cases. It aims to single out and address some of the main issues and problems that the application of this model raises, when applied to ordinary contexts, policies and design practices.

10.3.1 Spain

Although in recent years planning laws in Spain have paid more attention to excluding land from urban development, and to the presence in it of farmland areas, the statutory regime, classification and protection criteria of this land still remains weak and loose (Zazo Moratalla 2015a, cit., p. 2). Despite that, with the aim of protecting agriculture from the overwhelming wave of urban development—that took place after the return to democracy right up to the recent crash of the real estate bubble—peri-urban agricultural parks gained significant momentum in Spain, probably more than anywhere else in Europe. This was especially the case in Catalunya, where there are many real and working examples that can be seen.

It is not possible to outline the entire history of the Spanish AP in this chapter, due to the need for brevity. However, it is possible to summarize the main, common and meaningful aspects of the more significant ones, especially, the most popular and well-known: the *Park Agrari del Baix de Llobregat*. The parks that we are examining include the *Parc Agrari del Baix de Llobregat, de Sabadell, Parc Agricola y natural de Gallecs* and the ongoing proposition for the *Parc Agricola del Vallés.*[3] They are all under the administration of the *Catalunya Region* (*Generalidad*) and of the *Provincial administration of Diputaciò de Barcelona*, while the *Baix de Llobregat* and *Vallés* are also controlled by the *Metropolitan Government of Barcellona* itself.

[3]The Parc Agricola de Valles is still not formally operating. Despite that, the territorial dimension of Parc, the natural, farming and cultural values concerned and the complex web of socio-economic relationships active in the creation process, suggested to mention about the park main features.

Except for the *Baix de Llobregat* and the proposal for the *Parc Agricola del Vallés*—whose sizes amount respectively to 3,348 and 37,000 ha—the others share a mean dimension; *Sabadell* of 586 ha and *Gallecs* 733 ha. The majority of this land is used for agriculture in terms of either arable land or horticulture, although in the case of *Vallés*, the share of farmland slightly exceeds the amount of wooded land (respectively of 35% and 33%). The creation of each of these parks was brought about by a strong bottom-up commitment from peasants associations, environmental associations, social activists and citizens (in some cases supported by local municipalities), to hamper the urbanization pressure and farmland acquisition, that had been originating from the growth of the Barcelona core urban area. In the case of the *Parc Agrari del Baix de Llobregat,* a pivotal role was played by the participation, between 1996 and 1999, to the *LIFE European Programme*, that allowed the achievement of the first methodological, operational reference and financial tools—in collaboration with the Regional and Provincial bodies—to get started the first phase of park creation (Paül and McKenzie 2013). In the various parks discussed, and especially for the *Parc Agricola y Natural de Gallecs* and the *Parc Agricola de Vallés*—the environmental values of biodiversity enhancement, ecological connection and landscape improvement, the functions and role of active, rentable and ecologically oriented agriculture represent the main and underpinning concerns. This brings the conclusion that only viable peri-urban agriculture can compete with more profitable urban land use, and that the best way to achieve such a result is to develop stronger relationships with the urban market and citizens demand. This may come from either healthy and fresh food provisioning, or from ecosystem value and the deliverance of amenities.

In terms of land use, the protection of the AP area in the aforementioned cases is framed in the ordinary planning system, which range from the *Generalidad Regional Plan*, the *Diputaciò Provincial Plan,* up to the various *Municipal Pla d'Ordenacion.* Within these examples, the regime of park area protection is ensured or enhanced by the adoption of special and sectorial, regulatory and binding tools, directly managed by an AP partnership or consortium (Zazo Moratalla 2015b, cit., p. 75). In some cases, as for the *Parc Agrari y Natural de Gallecs*—encompassed by the Pla Regional d'espais d'interes Natural (PEIN)—the rules of protection are further fostered with the inclusion of the AP area in special environmental protection areas.

10.3.2 France

Likely due to its significant amount of open space in terms of cropland, pastures and wooded areas, the issues relating to the protection of rural areas observed in recent years in France has shown more evidence of attention on behalf of public bodies and policies (CGAAER and CGEDD 2009). This attention has been especially directed towards the enhancement and enforcement of a complete set of existing tools that, although issued at national level, asks for a closer commitment on behalf of local powers and authorities. This set of tools have allowed at local, and especially at inter-municipal level, both the protection of farmland and natural areas, and the defining of shared development and active management schemes, also in peri-urban contexts.

Examples of formal protection tools aimed at giving legal and binding protection status to natural and farmland areas, include the *Zones Agricoles Protégées* (ZAP) and the *Périmètres de Protection des Espaces Agricoles et Naturels Périurbains* (PAEN). The first can be issued with a Prefect decree in accordance with a request from the concerned municipalities, where the established protected perimeter can only be modified with the permission of the department and of its commissions. Similarly in terms of aims, the PAEN is issued on behalf of the responsible department in agreement with the municipalities concerned, and its aim is to affect the previsions of the local plan at the municipal and inter-municipal level (*Schéma de Cohérence Territoriale*—SCOT and *Plan Local d'Urbanisme*—PLU). Once stated, a PAEN perimeter can't be reduced or modified except with a Prefect decision. The interest of a PAEN rests in the fact that it is not limited to state a binding urban growth containment, but also to pair these measures with a development and enhancement action program (*programme d'action*) for the areas encompassed. Moreover, in the PAEN area and for its enforcement, some tools of land management are available on behalf of public authorities and aimed to foster public interest. They are represented by the *Societè de Amenagement Foncier* (SAFER) et by the *Etablissements public de Coopération Intercommunale* (EPCI), which are appointed, along with the bargaining authority, to compulsory or "friendly" purchase of natural and farmland areas of special interest, also resorting to pre-emption right. Within this regulatory framework, that in part provides some measures and opportunity related to the AP example, peri-urban farmland areas are affected by AP, or similar institutional and collaborative processes.

The first of these tools consists of the *Programmes Agriurbains* (Agri-urban Programs). These programs found their first formal acknowledgement in 2001 by the Comité Interministériel d'Aménagement et de Développment du Territoire (CIAD) and inspired the Peri-urban Area protection Act of 2005, which also established the PAEN tool (Poulot 2006). These Programs originated from 1967 with the *Plateau de Saclay* experience, that pioneered this kind of tool in the Ile de France (IdF). Here, an agreement was signed between the aforementioned SAFER and the *Agence pour la Protection des Espaces Verts*, aimed to protect some ponds and nearby farmland, and was managed at the inter-municipal level. Currently there are 10 formally established "programs" in the IdF, but others are in the process of development. Usually these agri-urban programs aim mainly to protect fragile natural areas and farmland close to urban environments. This starts with a bottom-up process of collective goal sharing between municipalities, farmers, not for profit associations and citizens/consumers. The area affected by the park is usually not narrowly established. Normally the program starts on a small scale, and graduates on to progressively larger ones, in relation to the development of a stronger vision between the partner participants, and of a "project territory" that best fits with local features and problems. The goals of these programs usually concern the protection of farmland and development of healthy and steady agriculture in close proximity to the city. In particular, they are aimed at developing shared and fair food production and provisioning schemes, as well as delivering amenities, environmental services and landscape quality, all features demanded by the urban population. In this sense, agri-urban programs aim to change the prospect of the separation and modern conflict between countryside, traditionally exploited as a resource by farmers, and countryside that is desired as a "living landscape" by citizens. This tries to contain the conflict accordingly with the new image of a «*nature countryside*» (*Campagne nature*) (Poulot 2006, p. 8). Food supply to the local proximity, landscape quality, and a healthy environment are the key elements pursued in the creation of such a new vision, where the countryside is conceived as "common" (*bien commun*).

Although agri-urban policies in France have developed with the aforementioned programs, and have also been inserted in the *Parcs Naturals Regionaux* policy, peri-urban Agricultural Parks are in some cases established and developed, with a strong stress on ecological features. In this context there are two important examples: The "Grand Parc" of *Miribel-Jonage of Lyon* and *Le Parc de La Deule of Lille*.

The development process of the *Gran Park of Miribel Jonage* started in 1968 with the protection of a mainly flooded area, close to the Rhone river course. The process originated from a top-down movement with the aim of protecting a sensitive environmental area, previously destined to an urban development. This area played a key role, in particular, as an area of groundwater recharging, which provided a drinking water reserve for the Lyon urban agglomeration. Progressively the park has grown and currently its area amounts to 2200 ha, of which 850 is woodland, 600 are protected areas (river and ponds, wetlands, shrubs, grasslands) and 400 is farmland (one third for organic agriculture). These areas are public owned, and for that reason, it was possible to create a contract with the local farmers to develop farming activities, which are aimed at providing a healthier way of cultivation and food production. Over the years the park has developed important functions as a leisure and landscape park, with a special focus on environmental protection and enhancement of biodiversity. Local farming activities are framed in this context. The general guidelines for management of the park were defined, from the onset, by a public-private partnership (*Syndacate Mixte*), which encompassed eleven municipalities, the Rhone Department and associated private bodies, whereas the operational activity was carried out on behalf of a Mixed Economy Society (*Societé d'Economie Mixte*).

The initial *Parc de La Deûle* project goes back to 1968 with a first proposal on behalf of *Organisme Régional d'Étude et d'Aménagement* (OREAM-Nord). The project aimed at the environmental recovery and rehabilitation of the industrial, extractive wastelands and brownfields fringes of the Lille south-west territory, and to the connection between Lille and the mines area of Lens along the Deûle canal. It encountered initial opposition, especially from farmers, and was finally re-launched only in 1993 on behalf of the *Inter-municipal Authority Lille Metropole*. The new project was finally established in 1995 and conceived from a public design competition. Its aim was the conciliation of strong environmental remediation of past industrial and logistic areas, and biodiversity enhancement, along with the support and regeneration of farming activities, which aimed to provide leisure services and food supply for urban needs. The triggering of the park project was eased by utilizing some public owned areas that were purchased with the prospect of the first project in the 1970s. The size of the park area is currently 400 ha, however the target was to reach up to 1000. As mentioned previously, this process and management model was strongly supported by public parties, and currently the park management is provided by the *Public inter-municipal partnership* ENLM (*Natural Space Lille Métropole*), created by the *Urban Community of Lille*, that recently became MEL (*European Metropolis of Lille*). This top-down approach had a strong economic commitment from public funding, with some 23 million euros allocated. The park implementation project has further witnessed strong involvement from farmers. After initial opposition, farmers have turned out to progressively become some of the main "allies" of the park development, thanks to a collaborative and participative approach adopted by the management authority. The agency *Natural Space Lille Metropole* has also been charged with technical management, relying on the support of a body of technicians transferred from the MEL staff. The park project has continued to progress, despite having to cope with ongoing urban development pressures which are still present, although reduced in recent years. From this point of view, the legal status of protection is provided by *Lille Metropole Schema de Coordination Territorial* (SCOT), the structural spatial planning tool, that unfortunately, doesn't have the power to define land use as binding on third parties; a prerogative appointed to the land use municipal plan (PLU).

10.3.3 Italy

Italy lacks, since many decades, of a suitable national law of reform of the actual normative for the spatial planning which goes back to more than seventy years ago. Despite some recent proposals, the national ruling framework in Italy doesn't contain a specific law addressing the issue of farmland protection, and acknowledging it as a specific matter. This domain is usually ruled by local ordinary planning tools, especially at municipal level. Some protection of peculiar natural/ rural areas, at least in terms of ecological and landscape values, is appointed to special planning tools mainly regionally issued, such as *Regional Landscape Plans* or Park and ecological Areas creation and management tools, drawing on national and European laws. This planning activity is usually enforced by the state at regional administrative level in collaboration with local administration.

Protection of "ordinary" farmland areas and AP promotion stemmed mainly from laws and tool issued at regional and local level.

In this context the *Parco Agricolo Sud Milano* is without doubt, the most significant example of an Agricultural Park. In particular, this is because of the pioneering role it has played either for its geographical dimensions and richness of natural and cultural endowments. Similarly to the other European cases that have been discussed, the first idea for a protected agricultural area in the southern region of the Milan metropolitan territory originated in the seventies, with an inter-municipal planning initiative. This action was progressively fed by the commitment of provincial and regional authorities, the involvement of citizens and by the proposition of "popular initiative laws" for the Agricultural Park in 1987. The final adoption of a regional law, that was issued in 1990, allowed finally the institution of the *Parco Agricolo Sud Milano.*

The area encompassed by the Milan South AP amounts to 47,000 ha, and includes the territories of 61 municipalities. The park contains a wide range of environmental, cultural, historical and agricultural features, and plays a key role in terms of both ecological connection and ecosystem services delivered to the southern sector of the metropolitan area. Agricultural activities in the park are mainly centered around cattle breeding for milk production and cultivation of arable crops, including rice. It has demonstrated a progressive shift towards organic food production and short supply chain practices aimed at feeding the city as well as park visitors.

The park is not endowed with a *Public/Private Partnership* (PPP) structure aimed at defining development directions and management activities. Such a function is appointed to a Directive Committee, constituted under the administrative authority of the Metropolitan City (Città Metropolitana), by the representatives of the mayors of the municipalities who have a direct interest in the park. Ordinary and operational management issues (e.g. advice on planning permission, technical support for farmers, park surveillance, etc.) are developed respectively by a Landscape Commission and by a Technical/agrarian Committee. From the point of view of the regulation of land use, the park—although formally defined since 1990 by a *Provincial Territorial Plan* (PTC)—is not directly managed by a park or metropolitan authority. In fact, the binding disciplines of spatial planning and land use, under the *Park Territorial Plan,* are administered by each individual municipality, to address matters concerning the park sector under their control.

The *Parco Agricolo della Piana Metropolitana* (Metropolitan Plain Agricultural Parc) in the Florence MA, as in the Milano South AP case, represents the end result of some prolonged studies and planning proposition activities. These activities were aimed from the outset, at creating a protected area on the outskirts of a very dynamic urban system, where a healthier and more viable urban environment could be fostered, especially by protecting farmland, landscape and ecological values. Although conceived in the 1960s the park—initially envisioned as an environmental/recreational one—found, in operational terms, a decisive phase of design and creation only since 2006. In this more recent period, the metropolitan park idea stemmed from an initiative from the Tuscany Region, that started a participative and innovative path in defining the design features and functional nature of a large, multi-functional, peri-urban park. This process allowed for a progressive shift from the initial idea of a mainly environmental park, towards a more complex project scheme for a multi-functional one, where the presence of sustainable agriculture was in synergy with the possibility of delivering leisure and ecosystem services. After a participative, and institutional consultation phase, the park project was finally adopted on behalf of the Tuscany Region in 2013 as a specific "Territory Project" of regional interest (*Progetto di Territorio di Interesse Regionale*), and inserted into the *Regional Territorial Plan* (PIT). The park project encompasses the territories of eight municipalities, and some 7,000 ha of farmland, wetlands and environmental regions. Its limits were precisely defined on behalf of the region, and had to be transposed—with limited changes—by the municipalities in their local plans. Despite its consultative process of creation, the *Parco Agricolo della Piana* was not supported by a specific administrative Agency, either private or PPP in nature, that aimed to promote and coordinate the park implementation and development phase. This turns out to be a significant weakness in terms of the park's stability, and implementation process, and therefore affecting ability to cope with the future. However, one improvement in this direction has come from a recent public call, issued from the Tuscany Region Agriculture sector in the framework of the *Rural Development Plan,* which is EC funded. This call has been aimed at co-funding *Integrated Territory Projects* (*Progetti Integrati Territoriali*), especially addressing environmental issues related to farming practices, and referring to specific territorial contexts, including the *Parco Agricolo della Piana* area.[4]

The City-countryside Park of Bologna Metropolitan Area (*Parco Città Campagna dell'Area metropolitana di Bologna*), is located in the south-western sector of the Bologna MA, and includes some 8000 ha of peri-urban open space. The target of the park's creation had been pursued from the beginning of the new century, with an inclusive process that was promoted by the Bologna Province and involved cultural associations, single farmers, citizens and concerned public bodies (Deriu 2008; Alampi 2010). This process was especially developed with the creation of a *Participative Urbanism laboratory* aimed to share ideas, problems and knowledge between the various social parts with an interest in the project. After 2006, the laboratory management passed to the *Metropolitan City of Bologna*, and then in 2010, the park area was formally established with an institutional planning agreement between the Metropolitan City and the other five municipalities, each containing portions of the park. From the outset, some of the goals of the park had been enhancement of landscape and cultural values, mainly deriving from the historical farming activities in the countryside. They also aimed to promote the recovery and support of multi-functional farming, suitable to deliver healthy food, ecological benefits, leisure and social services. In particular, the park project enhanced the network of slow countryside paths and helped create "food miles" schemes and locate farmer's market points.

As in the case of the Florence example, the park hadn't been endowed with an operative agency aimed at managing and implementing the park goals. However,

[4]The call for the project funding is issued by the Region, and open to public and private parties (e.g. farmers, farmers association, watershed authorithies, municipalities, etc.) proposing integrated projects referred to a specific territory.

spatial planning and land use issues usually have been managed by each municipality within their responsible area, taking into account the commitments and obligations assumed with their signing of the planning agreement.

10.4 Discussion

The analysis of a sample of regional cases of AP operational development, demonstrates some common features as well as peculiarities. This allows some key points to be stressed, in order to effectively appraise the critical issues that relate to the conception, design, launch phase and management of an AP.

The shifting concept of peri-urban areas and farmland protection

As shown by many examples, the genesis and original consolidation of the idea of an AP has frequently required a rather long "incubation" period. The creation process of many parks started twenty years or more before their first formalization. During this time-frame a conceptual bridge was often crossed, leading from a first "defensive" idea of protection of the countryside, especially related to distinctive ecological and environmental matters or values, passing to a more pro-active one. The latter has typically resulted in a vision where agriculture and farming activities played a pivotal role in looking after local natural endowments, delivering ecosystem services and promoting local food production. Included in this was often a vision of the park having a significant economic and developmental role.

Reframing the city-countryside relationships: some different accents

Although the importance of agriculture has clearly been acknowledged in each of the analyzed cases, farming activities and the presence of agriculture assume a more pronounced role in the examples of the Spanish AP, and, to a lesser extent, in Italy. In Spain, the continuity of peri-urban farmland areas and enhancement of agricultural systems (*dinamizacion*), have represented the starting point from which originated—as by-product—other AP's functions, services and practices. The idea of a "city-countryside pact", or of "countryside parks" (Donadieu 2006, p. 143), followed on from the French experience of agri-urban programs. Accordingly, the landscape oriented approach to farming followed in the French AP examples, tends to frame peri-urban agriculture mainly geared to answer new needs and demands coming the from city. This approach strongly relates to the idea of countryside conceived as "common" (Poulot 2006, p. 7), around which are stated new rules of consumption, production and use of territorial resources. It could be stressed that two such polarities, are usually not practiced in an alternative way and sometimes coexist together, although they can point out or result in different rethorics, action strategies, and project priorities.

The key role of governance dimension

As previously reported from Zazo Moratalla, the cases surveyed have confirmed the importance of a governance dimension in triggering AP creation, and their formalization processes. Even from the outset of initial AP creation, this should result in special attention paid to this matter. This is especially important in supporting areas of weakness in ordinary policy issues, such as farming, ecosystem services or farmers empowerment, and to better tackle conflicts of interests and differing visions related to them. Although this important aspect of governance in approaching AP creation and planning processes is shared between the different regional cases, in the Spanish examples it originates and is featured as a bottom-up movement. These movements primarily stem from issues of farmland preservation and protection to counter urban developments, mainly on behalf of farmers and environmental activists. Although in the examples of the French, and to a lesser extent in the Italian AP's a significant bottom-up movement was present, in the Spanish cases it represented a very strong feature that affected and directed the activity from the public bodies, either guiding the park vision and helping to formalize the AP management agency. From this point of view, it is possible to note how accurate and strong involvement or commitment from the social bodies (e.g. associations, activists, citizens, stakeholders) turns out to be of key relevance during the years of AP implementation.

An opportunity to innovate in Planning and design

The "biographies" of the different AP's examined, as well as others beyond the scope of this study, show how frequently the operational development of the concept and project of the AP, triggers a flow of ideas and a mutual exchange of institutional planning tools and policies. In particular, this relates to the general legal framework of reference—at national or regional level—concerning the acknowledgment of peri-urban areas peculiarity. In the French case, this relates to the nature and protection statement of peri-urban agricultural areas or, as in the Lombardy case, entails an extension of the concept of the Metropolitan Belt Park (*Parco Metropolitano di Cintura*). Beyond the legal toolbox, planning approach and design models are also concerned. From the planning point of view, practices related to AP creation and development raise the relevance and need for a multi-level and cross-sectorial coordination in public policy. In the aforementioned cases, regional, inter-municipal and municipal level are usually required to closely collaborate. In addition, in accordance with the goals and commitments of each national case, coordination is essential between the respective fields of urban development and design, infrastructure planning, environmental protection, rural development, and landscape planning. From this point of view the examples observed have proven how the conceived legal nature of the park institution and its status of autonomy, matter in improving the ease of such an integration. In fact, in the Spanish and French cases the formal creation of the park—jointly with legally defined boundaries, the presence of a PPP and a significant controlling role taken by the public partners—the AP implementation seems to better fit with a cross-sectorial approach and inter-institutional collaboration. This has apparently been more difficult in Lombardy and, especially in the Tuscan case, where the AP still strongly and primarily relates to the definition of a binding planning and land use tool, with only formal or weak connections with the policies of other fields, especially agricultural development, and without a defined program of ongoing and multilevel institutional collaboration. Moreover, the presence of a PPP structure seems to be a suitable tool to promote the adoption in fragile contexts—such as peri-urban farmland areas—of *Community Led Local Development* approach model (CLLD), that accordingly with the latest European Commission recommendations (EC 2014).

A territory design tool to connect policies and spatial planning innovation

Finally, as previously mentioned, the issue of AP spatial design deserves some consideration, especially for its potential in integrating and coordinating different policies, planning tools and actions. From this point of view, it is worth noting how the observable practices of the AP make room for a more accurate approach to landscape and territory design in peri-urban and farmland areas (Sabaté 2015). Tools of rural development usually relate only in a partial way to a systemic and territorialized vision of the possible impacts of rural development. This is because those measures are addressed to individual farmers, whereas the regional territory is conceived as an undifferentiated whole, and not as composed of different local rural systems or districts. Conversely, planning and land use planning tools are not allowed to question and regulate farming activities, except for some specific ecologically or environmentally sensitive areas. In this framework the role of the AP, especially in the design of spatial and environmental agro-ecological patterns can best match with the agro-environmental measures and *greening* regime, issued on behalf of the *Community Agricultural Policy*. In this sense the AP can improve the suitability and effectiveness of these measures and the achievement of the expected results.

This creates a more effective situation and a win-win relationship, which includes involvement from the farmers in the application of rural development and environmental protection goals. The agro-ecological and participative landscape design method experienced in the *Parc de La Deûle*, or the *Integrated Territorial Project* (PIT) recently proposed by the Tuscan Region for the *Parco Agricolo della Piana*, represents a step further in this innovative direction.

10.5 Final Remarks

The intertwined evolution of practices and theoretical reflections from the AP model—which has been synthetically designed—shows that it can represent an integrative and suitable planning and design tool that deals with the growing complexity of peri-urban open space governance issues. As seen in the context of European experiences, the AP has been used not so much as a protective tool for restricted environmental and valuable farming areas, but also as a device to treat with a pro-active approach the complex bundle of issues raised at the level of urban regional systems and underpinning new urban bioregions (Magnaghi and Fanfani 2009, cit.). In operational terms, the AP model embodies an opportunity to overcome the "business as usual" policy of distinctive separation between urban and rural domains. This jointly addresses spatial design issues and underpins schemes to create socio-economic place focused and networked partnerships, as well as local markets—either in a spatial or immaterial sense. It therefore stands to reason that it becomes possible to define the AP either as a "project of territory" or a "territory of project" (Magnaghi and Fanfani 2009, cit., p. 29), suitable to best match and integrate tools for rural development, urban policies and the features of an innovative peri-urban farming system. The cases reviewed have shown how the method of approaching and initiating the AP creation process, and the stability of the partnerships, are key factors in order to successfully implement the park project and goals. From this perspective, the bottom-up, inclusive approach and the involvement of social parties and citizens, who may even play a leading role, have turned out to be successful factors, whereas the only constraints observed through binding land use rules proved to be limited if adopted as only measure of protection. Different stresses in the framework of European cases and approaches to AP are paid to the role of farming activities and agricultural peri-urban systems. In this sense, although the multi-functional profile of agriculture in close proximity to urban influences embodies a pivotal role in reshaping some profitable form of farming activity on the city outskirts, it cannot be seen as an endorsement for a 'fictional role' in farming activity on behalf of farmers themselves (e.g. testimony of the lost values of the past civilization or supply of loisir services for the citizens). In this framework, profitable agriculture at the urban/rural interface and in regionally settled urban systems, needs to find in the AP tool pivotal support in the conciliation of new "urban countryside demands" from citizens, with a new "multi-purpose" role acknowledged to peri-urban agriculture. This entails new socio-spatial values to newly refuel and reshape societal relationships and delivering local and regional proximity—marketed and not—services. Finally, the AP seems pivotal in supporting the promotion and application of new design and integrative planning criteria and devices, that deal with the wide range of issues that stem from human settlements, and the global challenges that are at stake.

References

AA.VV. (2010). *Charter of peri-urban agriculture. For the conservation, planning, development and management of periurban agricultural spaces*. Castellfelds, Spain. https://areeweb.polito.it/didattica/UPWARD/dwd/agricolture/charter_periurban_agriculture.pdf. Accessed March 28, 2018.

Alampi, B. (Ed.). (2010). *Parco Città Campagna. La riscoperta della pianura bolognese*. Ferrara: Edisai.

Allen, A. (2003). Environmental planning and management of the periurban interface: Perspectives on an emerging field. *Environment and Urbanization, 15*, 135–147. http://eau.sagepub.com/content/15/1/135. Accessed February 9, 2015.

Allen, A. (2010). Neither rural nor urban: Service delivery options that work for peri-urban poor. In P. McCarney & M. Kurian (Eds.), *Peri-urban water and sanitation services. Policies, planning and method* (pp. 27–61). Berlin: Springer.

American Planning Association (APA). (2007). *APA policy guide on community and regional food planning*. https://www.planning.org/policy/guides/adopted/food.htm. Accessed March 7, 2018.

Bellamy Foster, J. (1999). Marx's theory of metabolic rift: Classical foundations for environmental sociology. *American Journal of Sociology, 105*(2), 366–405.
Brenner, N., & Schmid, C. (2015a). Towards a new epistemology of the urban? *City, 19*(2–3), 151–182.
Brenner, N., & Schmid, C. (2015b). Theses on urbanization. *Public Culture, 25*(1), 85–114.
Conseil général de l'Environnement du Développement durable (CGEDD), & Conseil général de l'agriculture de l'alimentation et des espaces ruraux (CGAAER). (2009). *Protéger les espaces agricoles et naturels face à l'étalement urbain.* http://agriculture.gouv.fr/ministere/proteger-les-espaces-agricoles-et-naturels-face-letalement-urbain. Accessed March 28, 2018.
Deriu, M. (2008). Il Parco Città Campagna Nella pianura bolognese. In D. Fanfani (Ed.), *Pianificare fra città e campagna. Scenari e progetti di nuova ruralità per il territorio di Prato* (pp. 137–151). Firenze: Firenze University Press.
Donadieu, P. (2005). Les Parc Agricoles urbains, Intérêts et limites. *Relation held for the Research Seminary, Il parco agricolo: esperienze italiane e francesi a confronto.* Empoli, Italy: Florence University (presentation not published).
Donadieu, P. (2006). *Campagna Urbane. Una nuova proposta di Paesaggio della città* (orig. title Campagnes Urbaines, Actes Sud/E.N.S.P., Arles, 1998).
ESPON. (2013). *EDORA, European development perspective in rural areas.* Applied research project, final draft. https://www.espon.eu/programme/projects/espon-2013/APplied-research/edora-european-development-opportunities-rural-areas. Accessed March 6, 2018.
European Commission (EC). (2014). *Community-led local development. Cohesion policy 2014–2020.* Bruxelles. http://ec.europa.eu/regional_policy/sources/docgener/informat/2014/community_en.pdf. Accessed March 28, 2018.
European Economic Social Committee (EESC). (2004). *Opinion on agriculture in periurban areas.* NAT/204. Brussels. September 16, 2004. https://www.eesc.europa.eu/en/our-work/opinions-information-reports/opinions/agriculture-peri-urban-areas. Accessed March 7, 2018.
European Region Committee. (2011). *Opinion of the committee of the regions on 'Local food systems'* (Outlook opinion). http://cor.europa.eu/en/activities/opinions/pages/opinion-factsheet.aspx?OpinionNumber=CDR%20341/2010. Accessed March 28, 2018.
Ferraresi, G., & Rossi, A. (1993). *Il parco come cura e coltura del territorio. Un percorso di ricerca sulla ipotesi del parco agricolo.* Brescia: Grafo.
Food and Agriculture Organization (FAO). (2015). *Urban and periurban agriculture.* http://www.fao.org/unfao/bodies/coag/coag15/x0076e.htm. Accessed March 07, 2018.
Gallent, N., Andersson, J., & Bianconi, M. (2006). *Planning on the edge. The context for planning at the rural-urban fringe.* London: Routledge.
Gree4Grey. (2016). *Green and blue infrastructure for grey peri-urban landscapes.* LIfe + Project. http://green4grey.be/en. Accessed March 07, 2018.
Hidding, M., Needham, B., & Wisserhof, J. (2000). Discourses of town and country. *Landscape and Urban Planning, 48,* 121–130.
Jacobs, J. (1984). *Cities and the wealth of nations. Principles of economic life.* New York: Random House.
Magnaghi, A. (2010). *Il Progetto Locale.* Torino: Bollati Boringhieri (IIth edition reviewed). English translation of first edition (2005) *The urban villages: A charter for democracy and local self-sustainable.* London: Zed Books.
Magnaghi, A. (2014). *La bioregion urbaine. Petit traité sur le territoir bien commun.* Paris: Eterotopia.
Magnaghi, A., & Fanfani, D. (Eds.). (2009). *Patto città campagna. Un progetto di bioregione urbana per la Toscana centrale.* Firenze: Alinea.
METREX. (2011). *Urban-rural relationships in metropolitan areas of influence: Best practice examples of metropolitan-rural cooperation.* Hamburg: Ministry of Economy, Transport & Innovation.
Millennium Ecosystem Assessment. (2005). *Ecosystems and human well-being: Synthesis.* Washington, DC: Island Press.

Montasell, J. P. (1996). *Els parcs agricoles. Concept, régim juridíc, òrgans I agents de gestió*. Diputació de Barcelona (note not published).
Morgan, K. (2009). Feeding the city: The challenge of urban food planning. *International Planning Studies, 4*(14), 341–348. https://doi.org/10.1080/13563471003642852.
Morgan, K., & Sonnino, R. (2010). The urban foodscape. World cities and the new food equation. *Cambridge Journal of Regions, Economy and Society, 3*(2), 209–224.
OECD. (2009). *Regional typology: Updated statistics* (pp. 240–253). Paris: OECD. www.oecd.org/gov/regional/statisticsindicators. Accessed February 10, 2014.
OECD. (2013). *Rural-urban partnership, an integrated approach to economic development*. Paris: OECD.
Paül, V., & McKenzie, F. H. (2013). Peri-urban farmland conservation and development of alternative food networks: Insights from a case-study area in metropolitan Barcelona (Catalonia, Spain). *Land Use Policy, 30*(1), 94–105.
PLUREL, Periurban Land Use Relationship. (2011). *Publishable final activity report*.
Polany, K. (1944). *The great transformation. The political and economic origins of our time*. Boston: Beacon Press (reprinted 2001).
Poulot, M. (2006). Les Programmes agri-urbains en Ile-de-Franc: de la "Fabrique" de territoires periurbains. *Communication presentée lors du Colloque de la dinamique des territoires ne milieu périurbain et le patrimoine naturel et culturel,* 26–27 Avril, Longueil, Université de Montréal.
Rururbal. (2011). Séminair Transnational: Les systèmes agroalimentaires locaux et les nouvelles politiques européennes. http://www.diba.cat/web/cjs/inici/-/cjs/1576. Accessed March 2, 2018.
Sabaté, J. (2015). *Reflexiones en torno al proyecto urbanistico de un Parque Agrario*. In C. Yacamán Ochoa & A. Zazo Moratalla (Eds.), *El parque agrario. Una figura de transitión hacia nuevo modelos de gobernanza territorial y alimentaria* (pp. 93–112). Madrid: Heliconia S. Coop.
Scott Cato, M. (2013). *The bioregional economy, land, liberty and the pursuit of happiness*. London: Routledge.
Shrijnen, P. M. (2000). Infrastructure networks and red-green patterns in city regions. *Landscape and Urban Planning, 48*, 191–204.
Stuiver, M. (2006). Highlighting the retro side of innovation and its potential for regime change in agriculture, between the local and the global: Confronting complexity in the contemporary agri-food sector research. *Rural Sociology and Development, 12,* 147–173.
Terres en Villes. (2008). *Guide de la Co-construction des politiques agricoles périurbaines d'agglomeration*. http://terresenvilles.org/wp-content/uploads/2016/11/TEV_CH1.2_GuideConstruction_2008.pdf. Accessed March 28, 2018.
Thayer, R., Jr. (2013). The world shrinks the world expands: Information, energy and relocalization. In E. Cook & J. J. Lara (Eds.), *Remaking metropolis* (pp. 39–59). Milton Park, Abingdon (UK): Routledge.
Tijallingii, S. (2000). Ecology on the edge: Landscape and ecology between town and country. *Landscape and Urban Planning, 48,* 103–119.
URBES/ICLEI Project. (2015). *Urban biodiversity and ecosystem services*. https://www.iucn.org/regions/europe/projects/cities-and-regions/urbes-project. Accessed 28/03/2018.
Van Leuven, E. (2010). *Rural-Urban integration; Town as focus point in rural development*. Dordrecht: Springer.
Viljoen, A., Bohn, K., & Howe J. (2005). *CPULS. Continous productive urban landscapes, designing urban agriculture for sustainable cities*. Oxford (UK): Elsevier.
Wandl, D. I., Nadin, V., Zonneveld, W., & Rooij, R. (2014). Beyond urban-rural classifications: Characterising and mapping territories in-between across Europe. *Landscape and Urban Planning, 130,* 50–63.

Zazo Moratalla, A. (2015a). *El Parque Agrario: estructura de preservación de los espacios agrarios en entornos urbanos en un momento de cambio global* (Ph.D. thesis). Escuela Politecnica de Madrid, Depart. de Urbanismo y Ordenacion Territorial, Madrid. http://oa.upm.es/39083/1ANA_ZAZO_MORATALLA.pdf, accessed July 2, 2018.

Zazo Moratalla, A. (2015b). Reflexiones sobre la protección de la base territorial del Parque Agrario: la institucionalización de su espacio agrario periurbano. In C. Yacamán Ochoa & A. Zazo Moratalla (Eds.), *El parque agrario. Una figura de transitión hacia nuevo modelos de gobernanza territorial y alimentaria* (pp. 73–92). Madrid: Heliconia S. Coop.

Zazo Moratalla, A., & Yacamán Ochoa, C. (2015). Introducción: estado de la cuestión de la figura de Parque Agrario. In C. Yacamán Ochoa & A. Zazo Moratalla (Eds.), *El parque agrario. Una figura de transitión hacia nuevo modelos de gobernanza territorial y alimentaria* (pp. 13–26). Madrid: Heliconia S. Coop.

David Fanfani is Associate Professor in Urban and Regional Planning at the Department of Architecture of Florence University and Director and lecturer at the Master degree Course in Regional Planning and design of Florence's Architecture School. His research activity focuses especially on analysis and design at the regional scale, addressing mainly matters related to periurban areas and reconnection between city and countryside. That accordingly with an integrated and cross-disciplinary bio-regional approach aimed to the recovery of a co-evolutionary relation between urban and rural domain. On these subjects he issued many publications and articles.

Chapter 11
Urban Agriculture in Urban Food Policies: Debate and Practices

Giacomo Pettenati

11.1 Introduction

With the implementation of institutionally driven strategic plans and with the engagement of food activists and actors of the food system, urban food strategies aim at planning and developing more sustainable, just, and resilient urban and city-region food systems (Sonnino 2009; Morgan 2015). Urban food production is often a key field of action of policies that often support and foster short food supply chains and environmentally and socially sustainable urban agriculture (UA). After an analysis of the role of UA in urban food systems and of its place in the theoretical debate about urban food planning, this contribution makes an analysis of six existing (even if not always already concretely implemented) urban food strategies in European and Northern American cities (Milan, Lyon, Brussels, Bristol, New York, Philadelphia). The main aim is to understand how these "new spaces of deliberation" (Moragues-Faus and Morgan 2015), can address UA and food system approach at the urban and metropolitan scale.

11.2 Urban Agriculture as Part of Urban Food Systems

Awareness of the urban nature of food related issues, which were once prominently confined to the rural sector, has gradually surfaced over the last 15 years. The reason for the absence of food from urban discourses (not from policies) are many: urban residents generally take food for granted; urban development in most of the

G. Pettenati (✉)
Dipartimento di Culture, Politica, Società, Università di Torino,
Lungo Dora Siena 100, Turin, Italy
e-mail: giacomo.pettenati@unito.it

E. Gottero (ed.), *Agrourbanism*, GeoJournal Library 124,
https://doi.org/10.1007/978-3-319-95576-6_11

Global North defines and plans cities as "non-agricultural" regions; technological changes in food processing and transportation, that allowed to feed urban residents with food coming from very far places; the progressive dichotomization of urban and rural policies (Pothukuchi and Kaufman 1999).

There are several reasons; however, why food should be considered an urban issue. First, the majority of the world's population lives in cities and, in some parts of the world, over 80% of the population is concentrated in urban spaces. Therefore, most consumers, whose individual choices are decisive in defining how the food system may evolve, are also concentrated in cities.

Cities are also the places where most decision-making powers are concentrated, thus directing the contemporary food system, governed by a few economic and political players that determine the characteristics of production, distribution and consumption at any scale (Morgan et al. 2006).

At the same time, cities are political and cultural arenas where movements of opposition and resistance to the dominant food system take place. The activism of *food movements* (Holt Giménez and Shattuck 2011), practices of *alternative food networks* (Goodman et al. 2012), and the increasingly widespread *urban food policy* initiatives (Moragues-Faus and Morgan 2015) will be discussed in the following paragraphs.

Nevertheless, cities are places where access to food can be problematic, especially in those neighbourhoods defined as *food deserts. Food deserts* are often characterized as places where it is very hard to find fresh and quality food, especially for culturally and economically disadvantaged subjects, or those with low physical mobility (Bedore 2010). Urban populations are also particularly vulnerable to the possibility of localised effects of some global dynamics (raw material prices, climate change, etc.) that affect food security (Morgan and Sonnino 2010). Urban ecosystems are very far from self-sufficient, cities largely depend on importations of food and the just-in-time supply system of the dominant urban retail model puts big cities "nine meals from anarchy" (Simms 2010). Should the flow of the food supply stop, then it has been calculated that there would be no more than three days of food autonomy for city dwellers.

The conceptual shift that characterizes the last 15–20 years of debate about the relationships between food and the city grounds on the gradual acknowledgement of the existence of an "urban food system", defined as the chain of activities related to the production, processing, distribution, consumption and post consumption of food (Pothukuchi and Kaufman 1999) including institutions and the related regulatory activities at the urban scale. Considering all food-related issues, networks, flows, problems and dynamics as part of a system, connected to other urban systems, reaches beyond the sectoral approach that is still typical of most urban food-related policies, moving toward a comprehensive approach that supports multi-dimensional and systemic "urban food policies".

Urban food production represents a fundamental element of the urban food system in terms of food security. In many cities of the *Global South*, commercial and household urban farming is one of the main food supply channels for less affluent residents (De Bon et al. 2010). Urban agriculture, in its different forms, though, is also becoming more and more politically relevant in cities of the *Global North*. The increasing visibility of urban agriculture is connected to the production of food for personal or commercial purposes, as well as to other motives related to community building, social inclusion, environmental education and so on (Ingersoll et al. 2008; Morgan 2015).

The potential role of urban agriculture in addressing the sustainability and the justice of urban food systems is acknowledged by policy makers and planners involved in the relatively new, but increasingly relevant, field of urban food planning. The next paragraphs discuss the role of the urban primary sector in the debate about *Urban Food Policies* (UFP)[1] by introducing a comparative analysis of six examples of cities of the *Global North*, focusing on how they approach urban agriculture.

11.3 The Emergence of Urban Food Strategies

First developed in, and still concentrated in, North American and British cities (Blay-Palmer 2009; Morgan and Sonnino 2010; Calori and Magarini 2015), UFP can generally be defined as voluntary policies that share many aspects with strategic planning including participated visions, integrated goals, mixed partnerships and active involvement of civil society (Calori et al. 2017). UFP are aimed at changing the city food systems (Moragues et al. 2013) and influencing the way food is produced, purchased, consumed and disposed of by urban dwellers.

Choosing a more explicitly political vision, we might say that the common thread of such policies «*is the political desire to address the socio-ecological problems associated with or generated by the industrial food system, including some or all of the following: planning barriers to urban agriculture, diet related diseases, carbon footprints, land conservation, food poverty, junk food clusters, urban–rural linkages and supermarket power*» (Morgan 2015, p. 10). Even if it is difficult to provide a complete account of existing, or still under development, examples of Urban Food Policies, a recent wide-scale study reported around 40 significant examples of such initiatives (officially approved UFP) worldwide, most of them concentrated in North America and Europe (Calori and Magarini 2015). Considering the signatory cities of the Milan Urban Food Policy Pact (an international agreement launched during the Milan Expo 2015), we might say that more than 160 cities all around the world are officially committed to work «*to develop sustainable food systems that are inclusive, resilient, safe and diverse, that provide healthy and affordable food to all people in a human rights-based framework, that*

[1]The strategic policies addressed to the urban food system are defined in different ways, such as urban food policies, urban food strategies, urban food visions, etc. In this chapter, I will always use the definition of Urban Food Policies (UFP), even if the analysed practices sometimes name them differently.

minimise waste and conserve biodiversity while adapting to and mitigating impacts of climate chang» (Milan Urban Food Policy Pact, art.1).[2]

Reviews and comparative studies about UFP have been produced over the years, with the aim of identifying common traits or transferable practices (Mansfield and Mendes 2013; Sonnino and Spayde 2014; Calori and Magarini 2015; Sonnino 2016; Calori et al. 2017).

Even if UFP have specific structures and objectives related to the geographical, social and political context where they develop and are implemented, we can identify some phases of the policy development process common to most of the existing experiences. The start-up phase can be informally initiated by the commitment of local interest groups (associations, fair trade economy networks, etc.). In other cases, it is directly institutionally trigged by local authorities. As the ones who have the formal legitimisation for approving and implementing food policies affecting the food system from the institutional and legal side, they are usually formally involved in the decision-making and policy-making process.

Once the process starts, an analytical phase usually follows, generally conducted by institutions, universities or other research centres, producing an assessment the food system. Often, simultaneously, a participatory process develops involving actors and stakeholders in defining the objectives and priorities of the future *food policy*.

The main outputs of these processes can be a vision statement, such as a food charter or a food agenda, that establishes the shared goals of future food policies.

In the most developed examples, when UFP are institutionally approved and implemented, this preliminary statement is usually followed by the adoption of a strategic document (e.g. a food plan), describing the goals of the food policies and the actions required to achieve them. The strategic document often suggests the establishment of a legally and politically legitimised management structure in charge for coordinating the implementation of the strategies and monitoring their achievements. This can be a specifically established new body such as a *Food Council* or a *Food Commission* or a pre-existing public or private actor of the urban food system (e.g. a food movement group) (Calori et al. 2017).

The initial context of development that triggers the process towards the definition and, sometimes, the adoption of urban food policies, varies. In North American cities, these initiatives moved from concerns related to public health (Morgan 2015) to an increased interest in socio-spatial justice in those urban areas characterized by the presence of *food deserts*.

In Europe, the landscape of urban food planning is more diverse. The most significant experiences take place in the United Kingdom, with an approach very similar to the USA and Canada. While in other countries, like Italy and France (Brand 2017), UFP development is based more on the active commitment of civil society and food movements. Recently, these movements seem to be evolving into more systematic approaches by involving institutions (Calori and Magarini 2015).

[2]*Milan Urban Food Policy Pact.* See: http://www.milanurbanfoodpolicypact.org (accessed 20/20/2018).

Cities of the Global South offer a different perspective where the core of food policies at the urban scale mostly focus on food security such as the pioneer case of Belo Horizonte (Rocha and Lessa 2009) or the experience of Johannesburg (Malan 2015). In most cases, from African and South American cities, the focus of such policies is on urban agriculture, which is considered a crucial method of increasing levels of food security in urban areas (Zezza and Tasciotti 2010).

Moragues et al. (2013) identify seven main thematic fields about which European UFP generally legislate, often with different approaches and objectives, while linked to local context: (a) *health and wellbeing* (e.g. improve the public health of urban population, increase the welfare services); (b) *environment* (reduce negative environmental impacts of the food system); (c) *economy and community development* (support local sustainable and just economy); (d) *social and cultural aspects* (support resilient communities, increase awareness about food system among urban population); (e) *food security/social justice* (improve access for affordable, culturally diverse and healthy food, guarantee a just food system); (f) *learning/empowerment*; (g) *urban-rural linkages* (foster the connections between city and the countryside through food).

This last point is strictly connected to the idea of pursuing a re-territorialization of food systems (Morgan et al. 2006). It is based on the re-localization of food networks and food supply chains coupled with a more complex vision of food consumption. This vision encompasses a territorial, social, and economic food system where production and consumption can no longer be disconnected. Furthermore, this same vision recognizes rural and urban areas as a part of the same regional system (Wiskerke 2009; Dansero et al. 2017).

As the next paragraph will describe, in detail, and with the support of case studies, urban agriculture, with its multidimensionality, is often seen as a fundamental step of this process of revitalization between food producers and food consumers. Urban agriculture is arguably an essential method for planning green spaces, a process aimed at reconnecting cities and their surroundings.

11.4 Urban Food Policies and Urban Agriculture: The Debate

Urban and periurban agriculture have always been components of urban food systems and of their landscapes and economies. Despite the appearance of agriculture and urbanism developing independently (Van der Schans and Wiskerke 2012), economic geography models clearly show the strict interrelationships between urban areas distribution and non-urban land use patterns (Sinclair 1967; Hite 1997).

According to Morgan (2015, p. 13), though, one of the hallmarks of the new urban foodscapes is *«the rapid growth of urban agriculture a broad definition which includes bee hives, urban chickens, front yard planting, community gardens, farmers' markets, green roofs and wild flower meadows among many other things, all of which are a stark contrast to the invisible and anonymous factory based foodscapes of the conventional food system».*

Contemporary[3] Urban Agriculture (UA) recently became central, again, in narratives and policies concerning the relationships between urban and rural areas (Donadieu 2006) and the development of urban food systems.

Originally, urban agriculture functioned to mitigate food security and provide a local food supply for people living in the city (Barthel and Isendahl 2013). However, urban agriculture is now complemented by other functions such as social, ecological, recreational, therapeutic, educational and aesthetic ones, whose importance is recognized both by institutional policies and by bottom-up practices (Kortright and Wakefield 2011).

The debate on urban food planning fully acknowledges the important role of UA as one of the main fields of interest of Urban Food Policies (Sonnino 2009; Wiskerke 2009; Morgan 2015). Before the striking emergence and diffusion of such debate, UA has mostly been approached by urban policies in the *Global North* with a sectoral perspective (mostly related to spatial planning or agricultural policies), which sometimes put it in conflict with other issues, such as public health regulations and land use rights. Considering UA as one of the dimensions of UFP may upscale the support to these activities, as they can be seen as a way to achieve bigger aims, like a more sustainable and equitable urban food system (De Zeeuw and Dubbeling 2009). Furthermore, considering UA as an effort to achieve a broader enhancement of the local food system can help to involve already existing food movements and community projects, which often consider UA as a way to increase the quality of individual and collective food access and to promote alternative models of urban living and economy (Tornaghi 2014).

The last consecration of UA as a pillar of UFP comes from the aforementioned *Milan Urban Food Policy Pact* (2015). Its premises express the commitment of signature cities in supporting it as a means for the development of more inclusive and sustainable food systems by acknowledging that *«urban and peri-urban agriculture offers opportunities to protect and integrate biodiversity into city region landscapes and food systems, thereby contributing to synergies across food and nutrition security, ecosystem services and human well-being»* (Preamble).

Despite the dominant role of UA in the debate about urban food planning, little research has been done about the actual approach of existing UFP in considering and supporting the practice of urban agriculture. Which scales of UA are considered

[3]I use the definition of "contemporary urban agriculture" to distinguish those practices of UA that emerged, especially in the Global North, after (sometimes as a reaction to) the global development of the agroindustrial capitalist food system (see Morgan et al. 2006; Wiskerke 2009), from those that historically locally provide food to cities. I want to highlight their nature of aware choice of development an unconventional (even if sometimes strictly institutionally regulated) multifunctional urban foodscape (Tornaghi 2014; McClintock 2014).

by UFP? Which dimensions of UA do they take into account? Which are the aims these policies try to achieve, through UA? How UA practices link to spatial planning and politics of urban space?

11.5 Which Scales and Dimensions of Urban Agriculture in Urban Food Policies?

In the rest of this chapter we try to answer to these general questions comparatively by analysing some examples of UFP that have been selected according to the priority they place on urban agriculture and the availability of their complete documentation online.

The source of information of this analysis are the "food plans" that have been produced by processes that are explicitly defined as *Urban Food Strategies* or *Urban Food Policies* (or with similar definitions) and that address the whole urban food system.

We excluded, therefore, some important cases of UA sectoral policies or projects, such as those of London or Detroit. This choice is due to the aim of this paper of exploring how institutional UFP deal with urban agriculture in discourse and actions. Of course, a complete understanding of urban food policies should also consider the role of some relevant actors of the food system (e.g. food movements, food policy councils, etc.) in contributing to the urban agenda food related issues, even without the formal approval and implementation of UFP.

The six urban experiences are concentrated in the *Global North* and notably in Europe (Bristol, Brussels, Milan and Lyon) and in the USA (Philadelphia, New York)[4] and have different times of development and implementation (Table 11.1). Some are pioneer experiences, like the case of Bristol that was the first city to establish a food council in the UK (Carey 2013), while some others are more recent. For example, Milan is still a work in progress: the guidelines for the food policy have been approved in 2015, during the EXPO (Calori et al. 2017).

As already mentioned, an analysis of UA practices and policies can be approached from different perspectives, coherently, with their multifunctionality (Sonnino 2009; Tornaghi 2014; Barthel et al. 2015; Hardman and Larkham 2014). Researchers may focus on the role of UA in increasing levels of food security or urban population, on its links to urban environmental quality, on community building strategies developed around them, on their place in urban planning and land use policies, or on their nature of social movements.

[4]UA in the cities of the Global South plays a different role in the urban food system, that deserves specific reflections, more directly connected with issues of food security and food sovereignty (Zezza and Tasciotti 2010; De Zeeuw et al. 2011).

Table 11.1 List of the analysed urban food policies

City	Year	Title	Country
Bristol	2013	A good food plan for Bristol	UK
Brussels	2015	Stratégie « Good food vers un système alimentaire durable en région de Bruxelles-Capitale	Belgium
Lyon	2015	Une alimentation Durable pour tous sur le territoire Lyonnais	France
Milan	2015	Linee di indirizzo della Food Policy di Milano 2015–2020	Italy
New York	2010	FoodWorks: A vision to improve NYC's food system	USA
Philadelphia	2011	Eating here. Greater Philadelphia's food system plan	USA

(*Source* Author's elaboration, 2017)

One of the approaches to a systematic analysis of urban agriculture practices can focus on their scales and scopes. Various classifications recognize at least three scales of such practices (Pearson et al. 2010; Mok et al. 2014):

- the microscale of punctual food production within the urban fabric (backyard gardens, roof gardens, balconies, etc.)
- the intermediate (meso)scale of community gardens and allotments in green urban areas
- the macroscale of urban and periurban commercial and professional agriculture.

This analysis uses scales of UA as entry points and conceptual frameworks for analysing how these activities are considered by UFP case studies (results are summarized by Table 11.2).

11.5.1 Microscale

The microscale of UA concerns food production happening in any possible useful space within the city: rooftops, school gardens, backyards, balconies, and so on. Only few of the analysed strategies explicitly put this punctual diffusion of agricultural food production among their principal aims. One example comes from the Lyon strategy, whose first objective is *«Produire partout où c'est (encore) possible [to produce food wherever it is still possible]»* (Ville de Lyon 2015, p. 7): rooftops, backyards, private and public unused green areas.

The main dimension of UA considered at this scale is probably the educational and cultural one. For example, the active involvement of school pupils in the development of a new food culture through school garden projects (Région de Bruxelles-Capitale 2015).

Table 11.2 Summary of scales, dimensions and some examples of proposed action, concerning urban agriculture in the analysed urban food policies

Scale	Dimensions	Actions
Micro (backyard gardens, roof gardens, balconies, etc.)	Educational Environmental Food security Community building	Development of school gardens (Bristol, Brussels, New York) Public support to the creation of rooftop gardens (New York, Brussels) Planning small community gardens in public green spaces (Lyon)
Meso (community gardens and allotments)	Educational Environmental Food security Community building Social inclusion	Creating and supporting community gardens (Bristol, New York, Lyon, Philadelphia) Promoting garden education (New York) Promoting access to garden allotments to low income families (Milan) Create a common framework of rules and aims (e.g. charta) of urban gardens (Milan, Lyon)
Macro (urban and periurban professional agriculture)	Food security Local economy Environmental Social inclusion	Producing knowledge about the state of UA (New York, Philadelphia) Protecting or creating key infrastructures for local food supply and short food supply chains (Bristol, Brussels, New York) Creating public markets (e.g. procurement) for locally produced food (Bristol, Brussels, Milano) Favouring access to land and to capital for local farmers (Milano, Philadelphia) Sustaining community supported agriculture (New York)

(*Source* Author's elaboration, 2017)

The New York City Council approaches UA with a similar strategy by identifying microscale UA using a multidimensional tool that *«offers significant opportunity to green our urban landscape, foster nutrition and food education, and help reconnect New Yorkers to their food»* (New York City Council 2010, p. 10). Very dense urban areas, like New York City, identify microscale UA—notably, rooftop agriculture—as the only way people living in central areas to reconnect, simultaneously, with food, with education, with the environment (greening roofs can contribute to a general enhancement of urban environmental quality), and, to some extent, mitigate food security. A first step in supporting such practices is identified in a census of rooftops potentially available for UA and a simplification of urban rules for their greening and ultimate cultivation (*ibidem*).

The Bristol strategy[5] proposes an interesting definition of the role of city dwellers as components of an *«innovative network of urban food*

[5]We consider here the more recent 2013 one, approved by the Bristol City Council, while a previous version was released by the Bristol Food Network in 2009.

producers» (Bristol City Council 2013, p. 17). They cultivate a wide range of urban spaces in order to contribute to achieving the objective of making the majority of urban population food growing within the next five years. The Bristol approach attributes a wider scope to microscale UA. The approach sees a *«potential to produce high cash-value products»* , which may contribute to the urban economy (ibidem).

11.5.2 Mesoscale

The intermediate scale of community gardens and individual allotments in urban green areas is probably the most representative of UA and is widely taken into consideration by the analysed food strategies.

Most food strategies identify this scale of UA as the most effective in increasing urban food security and involving a considerable sector of the urban population in food production: *«Community groups, enterprises and schools need to be supported to produce food on as much available land and in as many food growing spaces as possible»* (Bristol City Council 2013, p. 15).

The analysed strategies rarely provide detail regarding the specific activities and projects to be implemented, however, two main recurring typologies of UA practices are of this scale. They both aim to address food security, increase social inclusion, and invigorate neighbourhood revitalization.

The first type are individual allotments cultivated in urban areas for household consumption. They are one of the most common UA practice. Initially, they were informally developed in interstitial urban green spaces of many cities both in the *Global North*[6] and *South*. Progressively, they became institutionalized as objects of specific urban policies (for a brief history of UA see: Mougeot 2005; Hardman and Larkham 2014).

The access to allotments is often regulated according to income criteria, which favour low income households for access to food production in order to increase their food security and support their access to fresh quality food (see for examples the strategies of the Comune di Milano 2015).

The second typology are community collective gardens that are more explicitly linked to community and neighbourhood initiatives. Community collective gardens connect issues of household food security with strategies of social and cultural inclusion, neighbourhood revitalization, and diffusion of a new food culture. These goals are clearly stated by the *Greater Philadelphia Food System Plan* (DVRPC 2011), by the New York vision (New York City Council 2010), and by the Lyon

[6]An emblematic case is the one of the formerly industrial city of Turin (Italy), where contemporary UA start spreading in the 1960s in interstitial urban spaces—notably along the city riverbanks—mostly practiced by migrants attracted from other regions of Italy by jobs in FIAT car factories. In the last 30 years the creation and management of food producing allotments become one of the main actions of local institutions, joining social and environmental policies (Tecco et al. 2017).

strategy (Ville de Lyon 2015). Originally they were often conceived for their social functions, however, the inclusion of community gardens in the framework of UFP attributes a new meaning to them by linking them to more general reflections about urban food security and the need to rethink the categories of the "producing rural" and the "consuming urban" (*ibidem*).

11.5.3 Macroscale

Finally, macroscale UA, including professional and commercial urban and periurban agriculture, links urban food policies to traditional urban and regional spatial planning.

UFS deal with this type of UA from two main perspectives.

The first perspective addressing is local economy and its connections with the relocalization of the food system and the downsizing of food supply chains, often seen as a way for increasing the sustainability and the justice of local food systems, even if sometimes without an adequate critical perspective (Born and Purcell 2006). Many UFP focus on the potential of *«land based jobs»* (Bristol City Council 2013, p. 15), on the importance of safeguarding land for food production (*ibidem*), on guaranteeing affordable farmland around the city (Philadelphia: DVRPC 2011), on the development of a sustainable professional agriculture at the metropolitan scale (Région de Bruxelles-Capitale 2015), and on the need to use public agricultural areas for food production (Comune di Milano 2015).

The Greater Philadelphia strategy intimately explores issues related to macroscale (metropolitan) urban agriculture. The strategy identifies urban expansion pressures and land use and credit regulations as main threats to farmers accessing affordable land. The strategy proposes very detailed measures for addressing these threats. It states that food and farming, *«as cornerstones of a healthy regional economy, with adequate resources and support for business development and entrepreneurship»* (DVRPC 2011, p. 8), should be one of the pillars of a *«sustainable and resilient food system»* (*Ibidem*).

The second perspective relates to spatial planning and the role that UA can play in contrasting urban sprawl and helping the sustainable management of green areas around the city. UFP is distinguishable from previous, sectoral UA and spatial planning policies by its awareness of supporting urban and periurban agriculture. UFP is unique in its inclusion of land use and planning policies in addition to developing new planning and management approaches that address the whole food system, including productive urban and periurban landscapes. Urban and periurban agricultural land is thus considered a *«key infrastructure»* (Bristol City Council 2013, p. 21) of local food supply chains along with logistics facilities, markets, and so on.

11.6 Concluding Remarks

This analysis was based on the "food plans" (sometimes called differently) of the six case studies of UFP. This means that visions have been analysed. In order to comprehensively understand how UFP, as holistic and systemic policies aimed at developing more sustainable urban food systems, actually interact with UA, then it will be necessary to revisit these cities in few years with the purpose of investigating the implementation of these plans, monitoring their results, and analysing how they concretely affect UA local landscapes.

On the one hand, the recommendations of the UFP are based on some important potential leverages for their implementation. Notably, they require an increase of the general interest of city dwellers and consumers for locally produced food (Feldmann and Hamm 2015) and the general acknowledgment of the potential benefits of UA for urban society and the environment (Mougeot 2005).

On the other hand, though, the implementation of the UFP actions and recommendations for UA can be threatened by important barriers that characterize the agricultural system. First, urban land use policy, alone, cannot bypass strict regulations by supralocal policiesd or by market rules.

The real innovation of UFP in the UA debate and practice is to connect them systematically to broader discourses and polices, related to the sustainability and to the multidimensionality of the food system.

Some important issues, though, are missing or underrepresented in the analysed UFP experiences.

A critical approach to UA that is able to discuss its benefits while simultaneously facing and mitigating its controversial aspects is missing. For example, for what concerns: the potential socially exclusive and culturally exclusive access to its products (Tornaghi 2014), the pollution of urban natural resources used for food production (Brown and Jameton 2000), relationships between the public role of provider of affordable food and the *«neoliberal discourse of entrepreneurialism and self-help that shifts responsibility onto the shoulder of individuals and their communities»* (McClintock 2014, p. 154); consideration of other scales of food production as better than the local one for some supply chains (Born and Purcell 2006); analysis of the better alternatives to UA for land use in urban and periurban green areas (e.g. renaturalization, leisure, etc.) and of their potential integration with UA in the production of new urban landscapes (Viljoen and Howe 2012), within the framework of urban food policies (Pettenati 2017).

References

Barthel, S., & Isendahl, C. (2013). Urban gardens, agriculture, and water management: Sources of resilience for long-term food security in cities. *Ecological Economics, 86,* 224–234.

Barthel, S., Parker, J., & Ernstson, H. (2015). Food and green space in cities: A resilience lens on gardens and urban environmental movements. *Urban studies, 52*(7), 1321–1338.

Bedore, M. (2010). Just urban food systems: A new direction for food access and urban social justice. *Geography Compass, 4*(9), 1418–1432.
Blay-Palmer, A. (2009). The Canadian pioneer: The genesis of urban food policy in Toronto. *International Planning Studies, 14*(4), 401–416.
Born, B., & Purcell, M. (2006). Avoiding the local trap: Scale and food systems in planning research. *Journal of Planning Education and Research, 26*(2), 195–207.
Brand, C. (2017). Nascita e diffusione della questione alimentare urbana in Francia. *Bollettino della Società Geografica Italiana, 13*(10), 73–85.
Bristol City Council. (2013). *A good food plan for Bristol*. Bristol: Bristol City Council.
Brown, K. H., & Jameton, A. L. (2000). Public health implications of urban agriculture. *Journal of Public Health Policy, 21*(1), 20–39.
Bruxelles-Capitale, Region. (2015). *Stratégie Good food vers un système alimentaire durable en région de Bruxelles-Capitale*. Bruxelles: Region Bruxeles-Capitale.
Calori, A., & Magarini, A. (2015). *Food and the cities: politiche del cibo per città sostenibili*. Milano: Edizioni Ambiente.
Calori, A., Dansero, E., Pettenati, G., & Toldo, A. (2017). Urban food planning in Italian cities: A comparative analysis of the cases of Milan and Turin. *Agroecology and Sustainable Food Systems, 41*(8), 1026–1046.
Carey, J. (2013). Urban and community food strategies. The case of Bristol. *International Planning Studies, 18*(1), 111–128.
Comune di Milano. (2015). *Linee di indirizzo della Food Policy di Milano 2015–2020*. Milano: Comune di Milano.
Dansero, E., Pettenati, G., & Toldo, A. (2017). Il rapporto fra cibo e città e le politiche urbane del cibo: uno spazio per la geografia?, *Bollettino della Società Geografica Italiana, 13*(10), 5–22.
De Bon, H., Parrot, L., & Moustier, P. (2010). Sustainable urban agriculture in developing countries A review. *Agronomy for Sustainable Development, 30*(1), 21–32.
De Zeeuw, H., & Dubbeling, M. (2009). *Cities, food and agriculture: Challenges and the way forward*. Leusden: RUAF Foundation.
De Zeeuw, H., Van Veenhuizen, R., & Dubbeling, M. (2011). The role of urban agriculture in building resilient cities in developing countries. *The Journal of Agricultural Science, 149*(S1), 153–163.
Donadieu, P. (2006). *Campagne urbane. Una nuova proposta di paesaggio della città*. Roma: Donzelli.
DVRPC—Delaware Valley Regional Planning Commission. (2011). *Eating here. Greater Philadelphia's food system plan*.
Feldmann, C., & Hamm, U. (2015). Consumers' perceptions and preferences for local food: A review. *Food Quality and Preference, 40,* 152–164.
Goodman, D., DuPuis, E. M., & Goodman, M. K. (2012). *Alternative food networks: Knowledge, practice, and politics*. New York: Routledge.
Hardman, M., & Larkham, P. J. (2014). *Informal urban agriculture*. Berlin: Springer.
Hite, J. (1997). The Thunen model and the new economic geography as a paradigm for rural development policy. *Review of Agricultural Economics, 19*(2), 230–240.
Holt Giménez, E., & Shattuck, A. (2011). Food crises, food regimes and food movements: Rumblings of reform or tides of transformation? *The Journal of Peasant Studies, 38*(1), 109–144.
Ingersoll, R., Fucci, B., & Sassatelli, M. (2008). *Agricoltura urbana. Dagli orti spontanei all'agricivismo per la riqualificazione del paesaggio periurbano*, Bologna: Regione Emilia-Romagna.
Kortright, R., & Wakefield, S. (2011). Edible backyards: A qualitative study of household food growing and its contributions to food security. *Agriculture and Human Values, 28*(1), 39–53.
Malan, N. (2015). Urban farmers and urban agriculture in Johannesburg: Responding to the food resilience strategy. *Agrekon, 54*(2), 51–75.

Mansfield, B., & Mendes, W. (2013). Municipal food strategies and integrated approaches to urban agriculture: Exploring three cases from the global north. *International Planning Studies, 18*(1), 37–60.
McClintock, N. (2014). Radical, reformist, and garden-variety neoliberal: Coming to terms with urban agriculture's contradictions. *Local Environment, 19*(2), 147–171.
Mok, H. F., Williamson, V. G., Grove, J. R., Burry, K., Barker, S. F., & Hamilton, A. J. (2014). Strawberry fields forever? Urban agriculture in developed countries: A review. *Agronomy for Sustainable Development, 34*(1), 21–43.
Moragues, A., Morgan, K., Moschitz, H., Neimane, I., Nilsson, H., Pinto, M., et al. (2013) Urban food strategies: The rough guide to sustainable food systems. Document developed in the framework of the FP7 project FOODLINKS.
Moragues-Faus, A., & Morgan, K. (2015). Reframing the foodscape: The emergent world of urban food policy. *Environment and Planning A, 47*(7), 1558–1573.
Morgan, K. (2015). Nourishing the city: The rise of the urban food question in the Global North. *Urban Studies, 52*(8), 1379–1394.
Morgan, K., & Sonnino, R. (2010). The urban foodscape: World cities and the new food equation. *Cambridge Journal of Regions, Economy and Society, 3,* 209–224.
Morgan, K., Marsden, T., & Murdoch, J. (2006). *Worlds of food: Place, power, and provenance in the food chain.* Oxford: Oxford University Press.
Mougeot, L. (2005). *Agropolis: The social, political and environmental dimensions of urban agriculture*. London: Earthscan.
New York City Council. (2010). *FoodWorks. A vision to improve NYC's food system.* New York City: NYC Council.
Pearson, L. J., Pearson, L., & Pearson, C. J. (2010). Sustainable urban agriculture: Stocktake and opportunities. *International Journal of Agricultural Sustainability, 8*(1–2), 7–19.
Pettenati, G. (2017). Paesaggio e urban food planning: intersezioni teoriche e operative. *Bollettino della Società Geografica Italiana, 13*(10), 117–130.
Pothukuchi, K., & Kaufman, J. (1999). Placing the food system on the urban agenda: The role of municipal institutions in food systems planning. *Agriculture and Human Values, 16,* 213–224.
Rocha, C., & Lessa, I. (2009). Urban governance for food security: The alternative food system in Belo Horizonte Brazil. *International Planning Studies, 14*(4), 389–400.
Simms, A. (2010, January 11). Nine meals from anarchy. *The Guardian*, online version. https://www.theguardian.com/commentisfree/2010/jan/11/nine-meals-anarchy-sustainable-system. Accessed 8 March 2018.
Sinclair, R. (1967). Von Thünen and urban sprawl. *Annals of the Association of American Geographers, 57*(1), 72–87.
Sonnino, R. (2009). Feeding the city: Towards a new research and planning agenda. *International Planning Studies, 14*(4), 425–435.
Sonnino, R. (2016). The new geography of food security: Exploring the potential of urban food strategies. *The Geographical Journal, 182*(2), 190–200.
Sonnino, R., & Spayde, J. (2014). The new frontier? Urban strategies for food security and sustainability. In T. Marsden & A. Morley (Eds.), *Sustainable food systems: Building a new paradigm* (pp. 186–205). London: Earthscan.
Tecco, N., Coppola, F., Sottile, F., & Peano, C. (2017). Urban gardens and institutional fences: The case of communal gardens in Turin. *Future of Food: Journal on Food, Agriculture and Society, 5*(1), 70–78.
Tornaghi, C. (2014). Critical geography of urban agriculture. *Progress in Human Geography, 38*(4), 551–567.
Van der Schans, J., & Wiskerke, J. (2012). Urban agriculture in developed economies. In A. Viljoen & J. Wiskerke (Eds.), *Sustainable food planning: Evolving theory and practice* (pp. 243–258). Wageningen: Wageningen Academic Publishers.
Viljoen, A., & Howe, J. (Eds.). (2012). *Continuous productive urban landscapes.* London: Routledge.

Ville de Lyon. (2015). *Une alimentation durable pour tous sur le territoir lyonnais*. Lyon: Ville de Lyon.

Wiskerke, J. (2009). On places lost and places regained: Reflections on the alternative food geography and sustainable regional development. *International Planning Studies, 14*(4), 369–387.

Zezza, A., & Tasciotti, L. (2010). Urban agriculture, poverty, and food security: Empirical evidence from a sample of developing countries. *Food Policy, 35*(4), 265–273.

Giacomo Pettenati, Master in Geography at the University of Turin (2009), Ph.D. in Environment and Territory—Spatial Planning and Local Development at the Polytechnic of Turin (2014). He is currently post doc researcher in Political and Economic Geography at the Department of Cultures, Politics and Society of the University of Turin. At the same universities, he teaches Geography at the Department of Philosophy and Science Education and Food Geography and Spatial Analysis at the Master in Social and Environmental Sustainability of agri-food chains. His main current research fields are: food geographies, landscape management, mountain development, cultural heritage.

Chapter 12
Preserving and Restoring Italian Traditional Rural Landscapes. Addresses from Social Perception to Spatial Planning, Policies and Research

Bianca Maria Seardo

12.1 Traditional Rural Landscapes in Marginal Territories: Common Values and Threats

Rural landscapes enshrine great internal landscape diversity (Gambino 2009; Herrmann 2011; Sieber et al. 2010), contribute to regional landscape mosaics and public enjoyment and finally retain cultural, heritage, productive and economic potentials.

Antrop (1997, p. 109) defines traditional landscapes as: *«those landscapes having a distinct and recognisable structure which reflects clear relations between the composing elements and having a significance for natural, cultural or aesthetical values. Traditional landscapes [...] refer to these landscapes with a long history, which evolved slowly and where it took centuries to form a characteristic structure reflecting a harmonious integration of abiotic, biotic and cultural elements»* .

Traditional landscapes embed a number of both tangible and intangible values related to the delivery of cultural services such as leisure and recreational activities, learning—educational, calm and tranquility, escapism, sense of place (identity), sense of history or heritage, inspiration, spirituality (MEA 2005). Moreover both use-and existence values are recognized to traditional landscapes (Natural England 2009).

Under the 1992 *World Heritage Convention*, the *United Nations Educational, Scientific and Cultural Organization* (UNESCO) protects cultural landscapes across the world. According to the *World Heritage Convention* (UNESCO 1972), cultural

B. M. Seardo (✉)
DIST, Interuniversity Department of Regional and Urban Studies and Planning, Politecnico and Università di Torino, Viale Mattioli 39, 10125 Turin, Italy
e-mail: bianca.seardo@polito.it

E. Gottero (ed.), *Agrourbanism*, GeoJournal Library 124,
https://doi.org/10.1007/978-3-319-95576-6_12

landscapes represent the *«combined works of nature and of man»* (art.1) contributing to maintain or enhance sense of identity and spiritual values, but also biological diversity.[1] Following the Convention, cultural landscapes belong to three different categories: "clearly defined landscapes designed and created intentionally by man" (e.g. gardens and parklands), "associative cultural landscapes" and "organically evolved landscapes". This last category is particularly effective in describing traditional rural landscapes as resulting from *«an initial social, economic, administrative, and/or religious imperative and has developed its present form by association with and in response to its natural environment»* (UNESCO 2008, art.10-ii); they can be in the form of relicts (no more vital) or continuing—still vital landscapes (UNESCO 2012).

In the context of the Convention, traditional rural landscapes enshrine a very instrumental and actual value as they *«often reflect specific techniques of sustainable land-use [which can] contribute to modern techniques of sustainable land-use»* (UNESCO 2008, art.9). To date, the *World Heritage List*, encompasses as cultural landscapes 102 properties with 4 transboundary properties. Many of them are traditional rural landscapes.

Despite their high cultural and natural value, traditional rural landscapes are given a fragmentary attention possibly due to the progressive marginalization, fragmentation, abandonment, degradation, depopulation, obliteration of its cultural, landscaping and economic values. However common problems and threats can be identified (keeping in mind that, however, generalization is not the best method of investigation in the case of traditional rural landscapes).

Often, traditional rural landscapes reveal marginal territories. They are "marginal rural landscapes". Although the term "marginal" is perhaps somewhat disused, it seems to describe well some common conditions among these landscapes.

From the economical point of view, about one third of the European agricultural land is considered marginal (European Commission 2006 reported by Dwyer 2011).

Italy is at the 5th place among the EU countries for extension of rural land at risk of abandonment; at the same time, 33.4% of the "agricultural area" is a *High Nature Value* area (European Commission 2015). *High Nature Value* areas are intended as important areas for European biological and landscape diversity.

Dwyer (2011) characterizes marginal rural territories, identifying a series of connotative traits (common but also specific). There are natural constraints: poor soils, limited growing season, steep slopes, harsh climates. And also socio-economic constraints: geographically remote, poor infrastructure and communications; higher than average dependence upon primary sector, low incomes and low rate of employment. These conditions *may* lead to low standards of living and economic and demographic decline. On the other hand marginal rural territories usually encompass high rates of richness in cultural—and bio-diversity; they are favoured touristic destinations and production spaces with a proud population.

---

[1]See the 2014 UNESCO-SCBD Joint Conference on Joint Programme on Biological and Cultural Diversity: *Linking Biological and Cultural Diversity in Europe*, held 8–11 April, Florence, Italy.

The *International Terraced Landscapes Alliance*,[2] referring in particular to the marginal rural terraced landscapes, evokes a world-wide community of "terraced lands and populations" with common values and problems. This last concept combines the idea of marginality with a sense of community, conscience, connection, bond, which almost entirely opposes to the original meaning of the word. As always, a conceptual category if thoroughly studied can be questioned until it loses that solidity and compactness character conferred by its repeated use, until logically showing its more realistic complexity.

It has been known for a long time, but it is worthwhile to resume, the following concept well described by Guidicini (1986). He points out how the shift in the "technology" conception has played a relevant role in the growing marginalization of many rural areas. As "technologies" we can refer in a broad sense to the set of tools that humankind develops and applies for environmental control (thus also policies and programs can be intended as such).

The concept of technology, says Guidicini, has emptied more and more until it has lost its various and complex facets and has come to coincide with the idea of "machine". Technologies, thus reduced to pure mechanical elements, have increasingly grown out of specific environmental features: the machine by definition generates repeating actions, accentuates widely accepted situations, thus enhancing the role of centralities (the standard situation to which the process must adapt), leaving out marginalities. Environmental policies and intervention tools thus take on generic features when place-based solutions would be preferable. Complexity and diversification (such as the landscape diversity that characterizes much of the traditional rural landscapes) may come to be obstacles. This approach creates a depersonalisation of landscapes as interventions do not answer questions of diversity and complexity requested by traditional rural landscapes.

The value of site-specific intervention has now been recognized, but policies and approaches are difficult to follow. We may perhaps find this idea in the current place-based approach (Barca 2009) by virtue of which some EU policies are being re-thought.

12.2 Relevance of Population Consultation in Rural Policy-Making and Planning

The urbanized population increases. It will take only two centuries for most of the world's population to move from rural to urban: in the middle of the 19th century, 90% of people resided in rural areas, while it is expected that in the middle of the 21st century 70% will be urbanized (UN 2010). Moreover, the utilized agricultural area decreases.

[2]See: http://www.alpter.net (accessed 21/02/2018).

Why then to listen to farmers in marginal rural areas?

The *European Landscape Convention* states that the perception of the inhabitants must be taken into account within spatial policies and plans (CoE 2000). Namely, the Convention aims to *«establish procedures for the participation of the general public, local and regional authorities, and other parties with an interest in the definition and implementation of the landscape policies»* (art.5c). However, rural policies are often based on statistical collections, indirect analysis and consultation of accredited associations, rarely do they succeed to deeply survey the perception of local people and the cultural and social values expressed by the rural landscape. Moreover, *«the way people act is not completely determined by formal procedures or legal state regulations. Latent attitudes, informal social networks and practices»* (Volker 1997, p. 117) have to be investigated. Consultation and participation experiences on rural landscape carried out by planners are numerous, but often rely on the public discussion about single interventions on the landscape, not on the construction of large scale strategies (Cassatella and Seardo 2015).

Moreover, usually we tend to consider traditional rural landscapes as landslides where changes have taken place with less intensity and frequency, more or less completely survived the recent dynamics, so as to preserve the basic structure retained to this day. This is the general assumption of many policies and plans aimed at landscape conservation and management (Agnoletti 2010; Clark et al. 2004).

However, in some scientific arenas this way of thinking (in fact very useful for operational purposes) has been questioned as overly simplified (Renes 2015). Historical research, landscape ecology and other disciplines have contributed to demonstrate that the landscape is a dynamic entity and that its current structure is the peculiar result of not-linear stratification and transformation (Cevasco 2007).

Emilio Sereni (1961) reconstructing the history of the Italian agrarian landscape has been able to demonstrate, that it has been subjected to wide, not always slow, changes. A current "traditional rural landscape" has changed its basic layout several times over the centuries, as the case of expansion and retreat of the wooded areas around rural settlements, alternating to cultivated fields. It would be wondering which traditional rural landscape would we like to preserve today: that of the Middle Ages or the one of the Renaissance? Or the particular combination of the two which succeeded to reach the contemporary age (and therefore—as such—belonging only to our own time)?

Historical ecology shows that changes in traditional rural landscape can be more repentant and more usual than what one can superficially believe, demonstrating the intrinsic resilience of these kind of landscapes (Moreno 2003). Capable of continually transforming while maintaining some salient peculiarities. They also show that the visible layout of the territory that we name landscape is the product of tangible practices and intangible culture.

To answer the initial question: why to listen to farmers in traditional rural landscapes?

As seen, traditional rural landscapes would not actually be less affected by endogenous and exogenous transformation drivers, but rather they would ultimately have a greater resilience. Then robust policies and plans should deeply understand the factors of this resilience (social composition, individual motivation, local culture, quality of life…) (Dezio et al. 2016). Let just recall that anthropologist Ingold (1993) coined the term "taskscapes" emphasizing the importance of activities in the landscape rather than static ground/aesthetical patterns.

This means adopting a territorial approach to study traditional rural landscape.

12.3 Framing the Methodology: The Case Study of the Traditional Rural Landscape of the Morainic Amphitheatre of Ivrea

12.3.1 The Study Area

The study area is the *Morainic Amphitheatre of Ivrea in Piedmont* (Northern Italy), a remarkable natural amphitheatre of the Quaternary glaciations encompassing an area of 505 km^2, closed by a complex of lateral moraines (i.e. the Serra d'Ivrea) and encircling a 200 km^2 wide flat internal depression (Gianotti et al. 2015). We are in the geographical context that saw the development of the Olivetti company (in the first half of the 20th century), internationally known for the social role attributed to the factory.

The current spatial plans (Regional Territorial Plan, Regional Landscape Plan, Provincial/Metropolitan Territorial Plan) identify here a rural landscape of supra-local value due to the legibility of the traditional stratified farming systems (Roman centurization, hedge and tree-rows systems, extensive canal system, mountain and hill terraced slopes, diffused rural hamlets…). The inner landscape diversity is also a peculiar feature (Figs. 12.1 and 12.2). These values are ratified by extensive landscape protection areas, Natura 2000 sites, UNESCO sites and international recognition of the terraced vineyards cultural landscape.

This landscape is affected by nowadays common problems: abandonment of the steepest areas, depopulation and the rise of the average age in rural hamlets, increase of unmanaged natural dynamics that cause hydro-geological instability and obliteration of the cultural landscape, risk of landscape anonimyzation resulting from the expansion of monoculture. Moreover, there is perhaps a peculiar problem: as a vast and varied landscape, the population is reluctant to recognize it as a unitary one, thus not facilitating the implementation of appropriate coordinated policies and projects (Cassatella and Seardo 2015).

The macro-uses of the agricultural land are: arable crops 73%; pasture 13%, forest use 7%; vineyards 2%; poplars 2%; orchards 1%; other 1%; non-agricultural use 1%. Out of a total of 82 Municipalities and 125,000 inhabitants there are 1646 farms (Menegat and Perna 2011). 97% are individual farming-firms owing or cultivating an average of less than 10 ha each. The main economic technical orientation is arable land (50.51% of farms), but the second place is polyculture

Fig. 12.1 Traditional rural landscapes of the Morainic Amphiteatre. *Photo* Bruno Biava

Fig. 12.2 Traditional rural landscapes of the Morainic Amphiteatre of Ivrea (author's photo)

(13.51%), a sign of the still recognizable presence of traditional agricultural realities. The high average degree of mechanization reveals high fragmentation. However, many companies insist on difficult to mechanizable areas.

12.3.2 Aims and Methodology

Aim of the paper is to gather original elements to inform spatial plans and policies that have a particular impact on rural landscape, in order to better understand and thus preserve significant traditional marginal rural landscapes.

Such suggestions are developed through a case-study, following a bottom-up approach, based on the combination of farm-owners consultation and expert-led landscape analysis concerning a traditional rural landscape in Italy.

The methodology was developed by an interdisciplinary group consisting of researchers at the Politecnico di Torino,[3] following the request of the local Landscape Observatory (*Osservatorio del Paesaggio per l'Anfiteatro Morenico di Ivrea*[4]).

The team is constituted of spatial planners, architects, experts in forestry planning, and a sociologist.

The method consists of the qualitative approach of the focus-goup.

Each focus group lasts about two or three hours. It is held in closed doors, that is, without a public that could interfere with the free expression of ideas by the participants. Focus-goup sessions encompass about 10 participant at a time, so everyone can answer every question. The sociologist explains the motivations of the focus-goup and takes care of mediating the discussion, trying to ensure that it takes place following the scheduled order and to reach relevant responses (for further details, Seardo and Germano 2015).

The discussion schedule was agreed between the research group and the local Landscape Observatory. Below the trace of the questions submitted to the participating farmers:

- Do farmers think they play a role in preserving and transforming the landscape?
- Is landscape perceived as an income factor by farmers? Which current dynamics in the landscape have positive/negative impacts for farmers' business?
- What explicit suggestions should be addressed to planning and policy-making in order to preserve and actively safeguarding the marginal traditional rural landscape?

In 2013–2014, 4 focus groups took place, and about 40 people we interviewed.

---

[3]In particular: Prof. Claudia Cassatella and Prof. Alfredo Mela (Interuniversity Department of Regional and Urban Studies and Planning—DIST—Politecnico di Torino).

[4]See: https://www.facebook.com/osservatoriopaesaggioami (accessed 21/02/2018).

As one can see the sample is not yet a statistically relevant one, not comparable to the total population of the area nor to the number of farms. This is a first study to set up a more extensive and articulated methodology and later study.

All participants work in hill and mountain areas. Participants are wine-growers, livestock farms and mixed farms, as well as associations for the conversion of former abandoned lands into cultivated landscapes, trade associations, associations for the promotion of mountain productions.

The observation scale is the sub-metropolitan area, but it is analyzed on a more detailed scale looking at local actors and their perception.

This first fine-grained analysis is a methodological choice due to the fact that marginal rural landscapes easily escape from extensive area analysis and surveys.

12.4 Results

12.4.1 Farmers' Perception of Their Landscaping Action. Between Tenacious Pride and Partial Unconsciousness

In general, participants demonstrate awareness of their role in maintaining the traditional rural landscape.

All participants work in areas of difficult accessibility, low mechanization rates, inconsistency of local, national and international distribution networks for their products.

What is common to them is the will to resist in such territories, convinced of the potentialities and proud of their sense of identity. But not all of them are natives; then we will go also deeper into this. The main agricultural activities represented are hill and mountain livestock-farming, mountain viticulture, horticulture, collection and transformation of wild herbs, mountain trout breeding, agritourisms and small restaurants, curing-makers, cheese-makers, small mountain bakeries.

They point out potentials and obstacles of different nature.

First of all, some emphasize the importance of the entrepreneurial and social policies implemented in the first half of the 20th century by Adriano Olivetti: the company favored the hiring of many farmers in the factory but did not eradicated them from their own territories, also letting them to practice agriculture as second activity or in the free time. The effect of this far-sighted policy is recognized by the participants as one of the most important phenomena which took part in preserving the traditional landscape of the *Morainic Amphitheatre of Ivrea*.

The participants highlight the particular importance of mountain breeding in avoiding hydro-geological disasters, through the management of those relevant landscape elements that are not directly exploited for pasture purposes, such as beds and shores of watercourses, forest margins. These attitude is not common to all breeders, as some are more sensitive to the social utility of these interventions, others less.

Wine growers are perhaps more aware and proud of their landscaping action: according to them, maintaining terraced vineyards contributes to the beauty of the landscape and to the transmission of civil values such as labor and the meticulous knowledge of the territory in which they operate; they also hinder soil erosion and constructive knowledge, agricultural techniques and genetic varieties inherited from the elderly and adapted over the centuries for their fitness in these habitats.

Organic agriculture practices also emerge as important to maintain a valuable rural landscape: preserving soil quality, maintaining the elements of landscape which contribute to its diversification (forest-clumps, hedges and tree-rows, ponds…).

However, participants attending official certification schemes are the minority. Certification costs are likely to be added to ordinary running costs that are already higher than the those occurring in plain territories.

Beyond the purely agricultural practices, some feel the duty of divulge the values of the environment and the traditional knowledge. The enhancement is carried out both towards potential buyers and tourists, and towards the inhabitants themselves, since in many cases they are not aware of the territorial excellence they own.

What is said by regional plans to be the peculiar landscape value of this area (it is to say landscape diversification given by the presence of polyculture) is almost never explicitly invoked by the participants. That is, the choral value of each in shaping such a landscape seems to be undercut.

12.4.2 *Landscape as an Income Factor for Farmers? the Perceived Role of Disjoined Policies, Generational Discontinuity and New Farmers*

Participants agree that landscape could be an added value for their products. However, they say that there is still a lot to be done for this purpose: for the time being, there is no positive widespread or consolidated result in the area. Their products are still on the same market as those of larger companies supported by economies of scale. Landscape maintenance should be counted as an economic added-value facilitating the placement of products on different markets. In this sense, however, the following critical issues are pointed out: difficulties in accessing land and funding, lack of generational turnover, training needs, inconsistency of the promotion network.

Not a few people choose to return to the marginal and disadvantaged areas. Some return to manage the land of their respective families coming back from other professions, others approach agriculture for choice for the first time, others as a necessity due to the economic crisis. However, goodwill is not enough and finding financial resources is becoming more and more difficult (especially at this historic stage where even bank institutions do not easily grant liquidity): long and complicated administrative procedures for accessing public funds, high co-financing percentage required, are examples given by the respondents.

Getting into agriculture can be due to the lack of other job opportunities. Participants report several land reclamation initiatives for the start of small farms by unemployed people. Some young people solicited the administrations to help them engaging entrepreneurial activities on abandoned areas. This phenomenon also makes an interesting point, as these new farmers carry their previous skills (e.g. computer science) with them in trying to apply them in agriculture matters.

Less young interviewed often highlight the risk of the generational gap: many young people do not see gratifying prospects in farming and will not guarantee replacement in vineyards, orchards and alpine pastures. Anyone who will want to resume the cultivation of these landscapes will have enormous difficulties in the land recovery after years of abandonment. Someone launches the possible employment of the many refugees currently present on the territory to fill this generational gap. Table 12.1 synthesizes the main factors influencing the landscaping action of farmers, from their point of view.

12.4.3 Suggestions Addresses to Planning and Policy-Making Concerning Marginal Traditional Rural Landscape

The last question to participants concerned suggestions to planning and policy-making on how to maintain traditional rural landscapes.

The general hint by the participants in the focus-group points at the construction of a comprehensive and long-term project, difficult to find in the current territorial policies, in their view.

The respondents make more detailed proposals about this. The main are: education, new ways to access land and funding, active-protection of the landscape, place-based policies and regulations, supra-local coordination.

One pillar of this comprehensive territorial vision should be education. Education for farmers themselves, which sometimes lack the tools to convey the benefits brought by their activity to the community. Education about entrepreneurship and self-employment as, according to some respondents, it is preferable to policies distributing small scattered economic incentives on the territory. A local education system will be also useful to recover the phenomenon of school dropout in marginal rural areas.

Education is also seen as a means to spread awareness of the more minute aspects consolidating the aesthetics of traditional rural landscapes as traditional forms of vine cultivation which are gradually becoming more and more rare. Another "education" point should be the reduction of soil consumption in rural areas.

Table 12.1 Factors influencing the landscaping action of farmers (in the perception of the participants)

Driving forces and factors influencing the landscaping action of farmers	Positive perceived impacts on traditional rural landscape features	Negative perceived impacts on traditional rural landscape features
Former industrial policies	Factory did not eradicated farmers from their own territories. Agriculture as second activity. Landscape maintenance	
Actual spatial polices	Potential coordination with other sectoral policies: Enhancing landscape by revitalizing local economy	Disjoin of rural and landscape policies has come—for example—to risks of demolition of rural buildings Standardized interventions not suitable for mountainous landscape
Persisting individual values such as perseverance and proud of farmers in hill and mountain areas	Maintaining landscape also in difficult conditions Resilience Preserving soils from erosion Transmission of constructive knowledge, agricultural techniques and genetic varieties Will to communicate landscape and more general values	
Difficulty of land purchasing and renting	Potential support to new farmers, enhancement of traditional agriculture	Actually access to land and funding is very difficult
Physical accessibility to land	Development and maintenance of peculiar land exploitation techniques and morphologies (e.g. terraced landscapes) Mountain breeding and other practices of landscape-caring avoid hydro-geological disasters	Difficult mechanization, higher managing costs discourage young people, prospecting landscape abandonment
Generational gap	Potentiality of involving unemployed people and refugees in managing traditional rural landscape	One of the most evoked risk: abandonment of land and landscape
Organic farming	Soil quality preservation Maintaining landscape character (forest-clumps, hedges and tree-rows, ponds)	Adhesion costs to organic schemes maybe too high

(*Source* Author's own elaboration)

It is also required to identify more simplified and alternative forms of financing; those funds granted by the public Italian bank, the private sector, or private foundations that support new agricultural enterprises or existing ones through simplified procedures.

In the view of the respondents, the development of the rural area should be accompanied by regulations for the active protection of the landscape resource. In this sense, they propose to protect the fundamental natural resources at the very base of the agricultural activity: water, forests, wild products. Protection is to be intended as active-protection able to regulate the exploitation of the resource in order that it can continuously regenerate also thanks to a sustainable agricultural activity. This regulation should also protect local farming system from the appropriation by large companies of both material (wild herbs, genetic resources, water) and cultural resources such as traditional knowledge, according to Article 8(j) of the *International Convention on Biological Diversity*.

The participants need place-based policies and regulations for disadvantaged areas such as a specific law addressed to the mountain or more defined intervention measures for hill and mountain farms within the European rural development policy.

At the basis of all this, respondents would generally indicate the need of a supra-local coordination organism to overcome the excessive individuality of companies, to enhance farming-products and landscape promotion and to incentive join landscape management.

Finally, the need for the development of integrated policy and planning supporting traditional rural landscapes is outlined: recent territorial planning previsions seem to go to the detriment of the quality of the traditional landscape, establishing the new location of large facilities without adequate compensation measures.

12.5 Main Findings and Conclusions: Policies, Plans and Regulations Affecting Farmers' Landscaping Action

The study, though still on-going refers to a specific territory, though it aims to be a case by the means of which engaging broader debate in the European context.

As a matter of fact, the focus-groups raise many common issues for researchers and practitioners dealing with traditional rural landscape planning and managing such as: the role of farmers as landscape defense, the presence of high-value environmental assets, the fragmentation of land properties, the risk of generational gap, the difficulty of investment and the lack of alternative local food-chain supply, in addition to the intrinsic difficulties of hill and mountain farming. Finally, not all farmers are aware of the benefits that their activity brings to the community.

Overall conclusion based on focus-groups outputs seems to be that traditional rural landscape is at risk for several reasons.

One of the biggest is perhaps the fragmentation of policies and tools that should govern it. Despite many efforts being made at European, national and local level, farmers in the studied hill and mountain areas are uncertain about the future of their farms and of the landscape. The causes are a lot and should be investigated more deeply through interdisciplinary studies.

These critical points are often considered not directly dependent on spatial planning. Difficulties in purchasing or renting the land, generational gap, educational needs, inconsistency in the selling networks seem to be more connected to sectoral policies and regulations owned by agricultural and real-estate sectors.

However, if territorial planning is to support traditional and marginal rural landscapes, it must relate specifically to these factors intensely impacting on the rural landscape.

General addresses to policies and planning would be:

- Overcoming political disjoint. It is important to integrate spatial, rural and landscape policies not only at the strategic level but also at the implementation one (laws and regulations).
- As shown in Fig. 12.3 planning interest for rural landscape is increasing though its real capacity of influence is low. On the contrary, sectoral legislation (agricultural, sanitary, zootechnical, wine-growing sectoral laws plus organic schemes) have a relevant direct impact on rural landscape though not specifically taking into consideration landscape enhancement. Unwanted trade-offs may arise due to this condition.
- Connecting traditional practices and actual driving forces to landscape layout. Marginal traditional rural landscapes preserve live traditional knowledge about wise landscape management practices, so population and farm-owners consultation is fundamental for shaping sound landscape/rural policies and plans (Figs. 12.4 and 12.5).

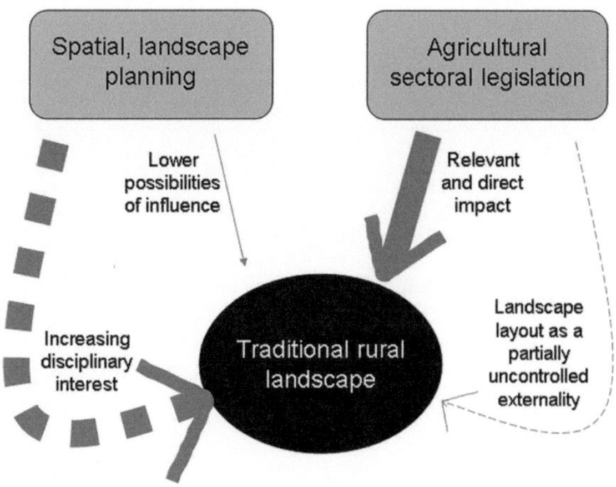

Fig. 12.3 Difference of impact of spatial planning and sectoral policies/regulations on traditional rural landscape. *Source* Author's own elaboration

Fig. 12.4 Traditional rural landscape and related still living practices of wise management. The manual mowing of mountain meadows allows the management of the spontaneous vegetation also along streams banks and fields margins (author's photo)

Fig. 12.5 Traditional rural landscape and related still living practices of wise management. Conveying the knowledge about traditional constructive methods allows farmers to maintain ancient landscape features of terraced vineyards, such as impressing dry stone walls—up to 4 meters high. *Photo* Xiangyi Tang

General addresses to research would be:

- Fine-grained analysis (both social and interdisciplinary-desk) is needed in shaping regional policies due to the high fragmentation of marginal traditional rural landscapes, which are more likely to go unnoticed by extensive-scale analysis.
- Research as a tool for the population. In fact, research is required by farmers—in this case—in supporting building a vision that neither farmers nor municipalities feel comfortable in doing, both for financial issues and internal conflicts.
- Finally, it is important to increase interdisciplinary research also involving disciplines pertaining to anthropology and possibly psychology: as a matter of fact the factors of personal motivation and sense of identity seem to play a relevant role in the resilience of these peculiar landscapes.

References

Agnoletti, M. (2010). *Paesaggio rurale. Strumenti per la pianificazione strategica*. Milano: Edagricole / Il Sole 24 Ore.

Antrop, M. (1997). The concept of traditional landscapes as a base for landscape evaluation and planning; The example of Flanders. *Landscape and Urban Planning, 38,* 105–117.

Barca, F. (2009). An agenda for a reformed cohesion policy. A place-based approach to meeting European Union challenges and expectations. In *Independent report prepared at the request of Danuta Hübner*. Commissioner for Regional Policy.

Cassatella, C., & Seardo, B. M. (2015). Cultural Ecosystem Services come strumento per la definizione di scenari e politiche di paesaggio partecipati. Una ricerca-azione nell'Anfiteatro Morenico di Ivrea. In *Proceeding of Conference XVIII Conferenza Nazionale SIU—Società Italiana degli Urbanisti ITALIA '45-'45. Radici, condizioni, prospettive*, Venezia (Italia) (pp. 452–459). Giugno 11–13, 2015.

Cevasco, R. (2007). *Memoria verde*. Parma, Diabasis: Nuovi spazi per la geografia.

Clark, J., Darlington, J., & Fairclough, G. (2004). *Using historic landscape characterisation*. English Heritage & Lancashire Country Council, Express Offset.

Council of Europe (CoE). (2000). *European landscape convention*. Treaty Series n. 176, Florence.

Dezio, C., Cavallo, A., & Marino, D. (2016). Resilient agrarian landscapes in face of changes: The coevolutive approach to understand the links between communities and environmental characters. In M. Agnoletti & F. Emanueli (Eds.), *Biocultural diversity in Europe* (pp. 339–351). Springer.

Dwyer, J. (2011). Marginal areas in rural Europe—Towards more appropriate policy support. In *Proceedings of the Conference: The Territorial Approach in Agricultural and Rural Policies. An International Review, Organized by INEA in Collaboration with RIMISP*—Centro Latino per el Desarollo Rural, Rome. November 4–5.

European Commission. (2006). *Scenar 2020. Scenario study on agriculture and the rural world*. Directorate General Agriculture and Rural Development, Directorate G. Economic analysis and evaluation, G. 4 Evaluation of measures applicable to agriculture; studies.

European Commission, Agriculture and Rural Development. (2015). *CAP context indicators 2014–2020*. 2015 update.

Gambino, R. (2009). Lectio Magistralis at Politecnico di Torino. http://www.cedppn.polito.it/content/download/350/1342/file/lectio_gambino.zip. Accessed 21 February 2018.

Gianotti, F., Forno, M. G., Ajassa, R., Cámara, F., Costa, E., Ferrando, S., et al. (2015). The ivrea morainic amphitheatre as a well preserved record of the quaternary climate variability (PROGEO-Piemonte Project, NW Italy). In G. Lollino, D. Giordan, C. Marunteanu, B. Christaras, I. Yoshinori, & C. Margottini (Eds.), *Engineering geology for society and territory* (Vol. 8). Cham: Springer.

Guidicini, P. (1986). *Il rurale riemergente. Un percorso storico su ipotesi di razionalità nell'agricolo e nella non città*. Milano: Franco Angeli.

Herrmann, S. (2011). Diversity of rural characteristics and its use for policy purposes in Europe. In *Proceedings of the Conference: The Territorial Approach in Agricultural and Rural Policies. An International Review,* organized by INEA in collaboration with RIMISP—Centro Latino per el Desarollo Rural, Rome. November 4–5.

Ingold, T. (1993). The temporality of the landscape. *World Archaeology, 25,* 152–174.

MEA - Millennium Ecosystem Assessment. (2005). *Ecosystems and human well-being: Synthesis*. Washington DC: Island Press.

Menegat, S., & Perna, N. (Eds.). (2011). *A.M.I. obiettivo primario. Il ruolo dell'agricoltura nell'Anfiteatro Morenico di Ivrea: produzioni, filiere e utilizzi del suolo*. Città di Castello: Nuovaprhomos.

Moreno, D. (2003). Activation practices, history of environmental resources and conservation. In G. Sanga & G. Ortalli (Eds.), *Nature knowledge: Ethnoscience, cognition and utility* (pp. 386–390). New York: Berghahn Books.

Natural England. (2009). *Experiencing landscapes: Capturing the cultural services and experiential qualities of landscape*. Natural England Commissioned Report NECR024, Sheffield.

Renes, H. (2015). Historic landscapes without history? A reconsideration of the concept of traditional landscapes. *Rural Landscapes: Society, Environment, History, 2*(1), 1–11.

Seardo, B., & Germano, R. (2015). Re-imagining landscape: The role of participation in landscape policy-making in the Anfiteatro Morenico di Ivrea. In *Proceedings of the UNISCAPE En Route seminar* (pp. 21–26), Torino. September 22–23, 2014.

Sereni, E. (1961). *Storia del paesaggio agrario italiano*. Bari: Laterza.

Sieber, S., Ratingero, T., Weingarten, P., Neumeier, S., Copus, A., Psaltopoulos, D., et al. (2010). *Building a tipology of european rural areas for the spatial impact assessment of policies* (TERA-SIAP). Joint Research Centre—Institute for Prospective Technological Studies, Luxembourg: Publications Office of the European Union.

UN. (2010). *World urbanisation prospects: The 2009 revision*. Department of Economic and Social Affairs—Population Division, New York.

UNESCO—United Nations Educational, Scientific and Cultural Organisation. (1972). *Convention concerning the protection of the world cultural and natural heritage*. Adopted by the General Conference at its seventeenth session. Paris. November 16, 1972.

UNESCO—United Nations Educational, Scientific and Cultural Organisation. (2012). *Operational guidelines for the implementation of the world heritage convention*. Intercommittee for the protection of the world cultural and natural heritage. World Heritage Centre 12/01—Annex n.3.

UNESCO—United Nations Educational, Scientific and Cultural Organization, Intergovernmental committee for the protection of the World Cultural and Natural Heritage. (2008). *Operational guidelines for the implementation of the world heritage convention*. Paris: UNESCO World Heritage Centre.

Volker, K. (1997). Local commitment for sustainable rural landscape development. *Agriculture, Ecosystems & Environment, 63,* 107–120.

Bianca Maria Seardo , Ph.D. in Regional and Urban Planning and Local Development is research fellow at Interuniversity Department of Regional and Urban Studies and Planning of Politecnico e Università di Torino. She is also mountain winegrower. Main research interests are landscape planning, especially concerning the connection of landscape research and policies to farming dynamics and sustainable practices and the development of landscape policies through farmers involvement.

Part IV
Agrourbanism Functions and Tools

Chapter 13
Towards Countryside Revival: Reducing Impacts of Urban Expansion on Land Benefits

Marco Marchetti, Matteo Vizzarri and Lorenzo Sallustio

13.1 Urban Expansion and Ecosystem Services Assessment

Ecosystem services (ES) are the benefits that people obtain from ecosystems, and are generated from the interactions between ecological structures and processes (including biodiversity[1]), and anthropogenic dynamics (social-economic frameworks) (Balvanera et al. 2017). Concepts around ES evolved since early 90s to explicitly recognize the role of nature (i.e., natural ecosystems) for human well-being, and directly quantify the human-induced effects (i.e., land use change; Foley 2005, and climate; Runting et al. 2017) on ecosystem health and functionality (Costanza et al. 1997; Costanza and Daly 1992). Hence, ES have simultaneously ecological, social and economic significance.

During the last years, several nomenclature systems have been proposed to best represent all ES, such as those from the *Millennium Ecosystem Assessment* (MEA

[1]Biodiversity is defined as: "the variability among living organisms from all sources including, inter alia, terrestrial, marine and other aquatic ecosystems and the ecological complexes of which they are part; this includes diversity within species, between species and of ecosystems" (United Nations 1992).

M. Marchetti · M. Vizzarri
Dipartimento di Bioscienze e Territorio, Università degli Studi del Molise, Con-trada Fonte Lappone, 86090 Pesche, Italy

L. Sallustio (✉)
Dipartimento di Bioscienze e Territorio, Centro di Ricerca per le Aree Interne e gli Appennini (ArIA), Università degli Studi del Molise, 86100 Campobasso, Italy
e-mail: lorenzo.sallustio@unimol.it

L. Sallustio
CREA Research Centre for Forestry and Wood, Viale Santa Margherita 80, 52100 Arezzo, Italy

E. Gottero (ed.), *Agrourbanism*, GeoJournal Library 124,
https://doi.org/10.1007/978-3-319-95576-6_13

2005) and *The Economics of Ecosystems and Biodiversity* (TEEB 2010). More recently, the *Common International Classification of Ecosystem Services* (CICES) has been adopted in Europe (Haines-Young and Potschin 2013). Differently from previous frameworks, CICES recognizes biodiversity as an intermediate service, instead that a final service itself. Indeed, various components of biodiversity constitute the ecosystem's structure, and regulate both state and rate of many ecological processes, being a support for producing and delivering other services (Cardinale et al. 2012; Mace et al. 2012). CICES is based on the following ES categories: provisioning, regulation and maintenance, and cultural. Provisioning services are directly obtained from ecosystems and regard raw materials and other "tangible" goods, such as crops, freshwater, fruits and berries, and timber. Regulation and maintenance services derive from the stabilization and control roles by many ecosystem processes, such as on physical, chemical, and biological conditions (both in atmosphere and soil). Cultural services are considered "intangible", as they reflect the human-nature relationships, such as sense of place, identity, aesthetic values, historical and religious features, etc.

ES are in fact strictly linked to the way through which humans interact with nature, and subsequently which products, goods and services, that ecosystems are able to produce and deliver. Thus, approaches to ecosystem stewardship passed so far throughout sustainability, adaptation and resilience, which all define the trajectories to maintain functionality of nature and its components for preserving the wellbeing of human generations over the time. Recently, both sustainable development goals and planetary boundaries represented a two-side condition, as on one hand targets for sustainability are set (Griggs et al. 2013), and on the other hand, thresholds for development are expected to be taken into account (Steffen et al. 2015). On the policy side, several regulations and directives have been implemented during the last decades to ensure that biodiversity and ES are maintained and preserved now as well as in the future, such as the *UN Convention on Biological Diversity* and the *EU Biodiversity Strategy*, at global and European scale, respectively.

According to sustainability purposes, policy-makers, scientists and administrations in general, continuously face with challenges regarding the ES assessment. ES are assessed by following a demand-supply chain, especially through using indicators or proxies, and in both biophysical and monetary terms (e.g., Layke et al. 2012). Available modelling and mapping approaches are generally oriented to deeper understand stocks and flows of ES at multiple spatial and temporal scales (Bagstad et al. 2013; Burkhard et al. 2013; Palomo et al. 2017). One of the crucial points while assessing ES is to make their respective benefits explicit in monetary units (i.e., Total Economic Value; Brander and Crossman 2017), and in turn identify the best balance among different ES values within the same portion of territory (i.e. quantifying ES trade-offs; Rodríguez et al. 2006). Mapping approaches allow to visualize ES stocks, monitor possible changes in ES fluxes, and highlight peculiar ecosystem features, such as "hot spots" and "cold spots" of ES supply and demand (i.e., Sallustio et al. 2017a). Through simulating future landscape evolutions (i.e., scenarios), ES stocks and flows can also be mapped at different temporal steps. As a consequence, policy-makers and spatial planners can use ES maps to orient decisions towards e.g., maximizing services for a certain area, or balance them in a given region, or for different time periods, and thus ultimately support the implementation of sustainable purposes for land governance (de Groot et al. 2010).

The increasing climate and land use changes (LUC) offer evidence that human generations entered in the Anthropocene era, which describes the profound anthropogenic modifications of the biosphere (Ellis et al. 2010). During the last Centuries, one of the most critical anthropogenic signatures on Earth has been revealed through LUC and subsequent land degradation phenomena. More in depth, two main interconnected phenomena of land transformation have been detected at global scale, such as forest transition (Meyfroidt and Lambin 2011) and urban growth (Grimm et al. 2008). All of them cause changes in ecosystem health and functionality, and in turn in ES stock and flows, thus modifying land benefits for local communities and having negative impacts on economic and social systems as well (e.g., Foley et al. 2005).

During the 2000–2006 period, the most relevant LUC in EU have been the increase of artificial areas (+3.4%, considering both urban areas and infrastructures) and forests (+0.1%), against the decrease of pastures and grasslands (−0.3%) as well as croplands (−0.2%), followed by a slight decrease even in seminatural vegetation and wetlands (EEA and JRC 2010). More in depth, from 1990 to 2006 artificial surface increased by 8.8% in EU, of which approximately 51% is considered as sealed (Stolte et al. 2016).

In most cases, urban expansion is not strictly related to new household demand. This is demonstrated by the fact that urban areas and infrastructures are still expanding, even where human population is decreasing or at least not growing (e.g. some regions in Italy; ISPRA, 2016, and East Germany, Haase et al. 2013). This is mostly due to the so-called "urban shrinking" (Haase et al. 2014), which is considered an emerging issue in Europe (Kabisch and Haase 2011). Particularly, shrinking cities generally imply the decrease of population within existing urban areas and new low-density settlements outside (Salvati et al. 2012). Urban expansion and subsequent land take, particularly in shrinking contexts, have high social, environmental and economic impacts, and in turn undermine human well-being and sustainability in urban areas and their surroundings. Additionally, it is important to highlight that other than the magnitude, the spatial configuration of the new low-density settlements, usually referred as "urban sprawl" (EEA 2006), further exacerbates the negative effects of urban expansion on ES. This is particularly true in Countries in the Iberian, Mediterranean and Balkan areas, where an alternative definition of the form of urban growth has been recently proposed by Romano et al. (2017). Focusing on Italy, the authors proposed the so-called "urban sprinkling" model, literally *« a small quantity falling in scattered drops or particles »* (Romano et al. 2017, p. 1), which is characterized by highly fragmented and scattered settlements with higher irreversible environmental impacts. Such phenomenon is more difficult to be controlled and mitigated than urban sprawl (Hennig et al. 2015).

With particular regard to Italy, the high magnitude of the phenomena (+30 ha per day of artificial areas from 2016 to 2017)—one of the highest in Europe in spite of the economic crisis and the stable demographic trend (ISPRA 2017), in conjunction with its peculiar spatial configuration (i.e., sprinkling)—highlights the urgent need to develop new strategies and policies able to curb urban expansion. At the EU level, the urgent need of monitoring and halt urban growth as well as its consequences (especially soil erosion and the loss of soil organic matter) represents a major environmental challenge (see Paleari 2017, for a detailed review on the EU legislation framework). In particular, it has been proposed to have policies targeting no (zero) net land uptake by 2050 (European Commission 2011). In spite of such ambitious target, an EU soil-specific legislation is still missing, and the roadmap towards stopping urban expansion mostly disregarded (Paleari 2017). Again, the Italian case appears emblematic in this sense. In fact, after almost four years of debate, a specific law to halt urban expansion is still unapproved, and the existing regional laws are often heterogeneous, weak and poorly applied (ISPRA 2017). As a consequence, the evaluation of both ecological and economic impacts on ES from urban expansion, represents a tangible instrument to (*i*) increase the awareness and of the public opinion and, in turn, of policy-makers, and (*ii*) complete the costs-benefits analysis related to policies, plans and projects with environmental aspects otherwise neglected (Jansson 2013; Sallustio et al. 2015).

In order to further understand in which way policy and science currently address the linkages between urban expansion and ES provision, we carried out a literature review based on SCOPUS[2] database. The review exercise aims at describing the state and trend of current research focusing on the effects of urbanization on ES provision due to land take and soil sealing processes. In addition, the review exercise tries to provide scientific basis and robust examples on how current ES assessment methods are adopted for and may improve the evaluation of urbanization effects on land benefits in economic terms. The review exercise considers peer-reviewed articles published from 1960 to 2017; it is based on a combination of keywords (i.e., research fields), and it is downscaled (from global to European to Italian). For the purposes of the review exercise, soil sealing and land take are both considered effects of urban expansion (cf. European Commission 2012).

Table 13.1 reports results per target-phenomenon, adopted search strengths, keywords used, including their combination at different scales. Selected contents from review results have been then deeply analyzed and discussed according to their relevance and consistency with the "urban expansion—economic effects on ES (i.e. land benefits)" topic.

Main findings from the review exercise reveal that: (*i*) the number of records strongly reduces from global to national scale (75–95% of difference between the two scales for all search strengths), and is particularly relevant in Italy (4–20% of the total number of records for all search strengths); (*ii*) papers treating the economics of ES are less than those concerning ES in general (38% of total records);

[2]See: https://www.scopus.com/search (Accessed: 22/02/2018).

Table 13.1 List of search strengths and combined keywords as adopted for the review exercise, grouped by target-phenomenon and geographic scale

Phenomenon	Scale	Search strength(s)	Records found
Urban growth effects on ES	Globe	"Urban growth" OR "urban expansion" OR "urban sprawl" AND "ecosystem service*" OR "environmental service*"	306
	Europe	"Urban growth" OR "urban expansion" OR "urban sprawl" AND "ecosystem service*" OR "environmental service*" AND "Europe"	15
	Italy	"Urban growth" OR "urban expansion" OR "urban sprawl" AND "ecosystem service*" OR "environmental service*" AND "Italy"	16
Detailed effects of urban growth on ES (namely, soil sealing and land take)	Globe	"Soil sealing" AND "ecosystem service*" OR "environmental service*"	19
	Europe	"Soil sealing" AND "ecosystem service*" OR "environmental service*" AND "Europe"	6
	Italy	"Soil sealing" AND "ecosystem service*" OR "environmental service*" AND "Italy"	5
	Globe	"Land take" OR "land uptake" AND "ecosystem service*" OR "environmental service*"	11
	Europe	Land take" AND "ecosystem service*" OR "environmental service*" AND "Europe"	1
	Italy	"Land take" AND "ecosystem service*" OR "environmental service*" AND "Italy"	2
Urban growth effects on ES-economics	Globe	"Urban growth" OR "urban expansion" OR "urban sprawl" AND "ecosystem service*" OR "environmental service*" AND "economic*"	111
	Europe	"Urban growth" OR "urban expansion" OR "urban sprawl" AND "ecosystem service*" OR "environmental service*" AND "economic*" AND "Europe"	7
	Italy	"Urban growth" OR "urban expansion" OR "urban sprawl" AND "ecosystem service*" OR	9

(continued)

Table 13.1 (continued)

Phenomenon	Scale	Search strength(s)	Records found
		"environmental service*" AND "economic*" AND "Italy"	
Detailed effects of urban growth on ES-economics (namely, soil sealing and land take)	Globe	"Soil sealing" AND "ecosystem service*" OR "environmental service*" AND "economic*"	9
	Europe	"Soil sealing" AND "ecosystem service*" OR "environmental service*" AND "economic*" AND "Europe"	3
	Italy	"Soil sealing" AND "ecosystem service*" OR "environmental service*" AND "economic*" AND "Italy"	1
	Globe	"Land take" OR "land uptake" AND "ecosystem service*" OR "environmental service*"1 AND "economic*"	4
	Europe	"Land take" AND "ecosystem service*" OR "environmental service*" AND "economic*" AND "Europe"	0
	Italy	"Land take" AND "ecosystem service*" OR "environmental service*" AND "economic*" AND "Italy"	1

Number of records found per each search strength is also reported
* Indicates that the review has been conducted for both singular and plural forms
Source Author's elaboration

and (*iii*) records referring to land take are less than those focusing on soil sealing as urban growth effect. No result was obtained when considering land take impacts on ES-economics at European scale. A limitation of the review exercise and comparison among records may be related to the return overlapping records at multiple scale of analysis (no exclusivity).

From global to national scale, most relevant results present different methods and approaches for assessing ES in urban-rural interface. Not all of them address the ES assessment in both monetary and spatially-explicit terms (e.g., energy supply-demand in Germany; Kroll et al. 2012; interactions between ES and urbanization problems in China; Li et al. 2016). When downscaling the review exercise, contents from several records demonstrate that urban growth and more generally urbanization imply a loss of ES. In Mexico, Mendoza-González et al. (2012) argue that urban sprawl in conjunction with the reduction of mangroves and coastal dunes originated a loss of the total ES value (averagely, 2.9 million USD) from 1995 to 2006. In China, Zhang et al. (2015) detected an increase of artificial surfaces mainly at the expense of cultivated lands, which in turn originated a loss of the ES value up to approximately 2.3 billion Yuan from 2000 to 2010. In Italy, Fedrigotti et al. (2016) highlight that urban expansion over grasslands and croplands partially triggers a reduction of the economic contribution of ES during a 152-years period. In Switzerland, Grêt-Regamey et al. (2008) simulated the impact of LUC scenarios on several ES, and found that urban expansion contributed to a loss of 1.9 million CHF for scenic beauty benefits by 2050. According to the main review findings, we can generally argue that: (*i*) the effects of urban expansion on land benefits are directly dependent upon the economic value of the lost land use (e.g., wetland, grassland) (Aretano et al. 2013; Zhang et al. 2015); (*ii*) changes in land use intensity have higher impact on ES supply than changes in land cover (Kroll et al. 2012); and (*iii*) not always urban expansion leads to a loss of ES values when e.g., the inclusion of natural areas (e.g., city parks) in urban contexts is considered during land planning processes (Mendoza-González et al. 2012).

When detailing the review, we found that “land take” is more difficult to isolate as a direct urban expansion effect on land benefits, in comparison with “soil sealing” (see comments above). While “land take” refers to the increasing of artificial areas over the time, “soil sealing” is the “permanent covering of soil with impermeable material” (European Commission 2012). In light of this, some parts of land take will results in soil sealing (Gardi 2017), where the latter is more specifically considered to further understand the direct and specific impacts of urban expansion on ES provision, and in general on land benefits (Artmann 2016).

A considerable portion of records refers to the methodology adopted to evaluate land use change dynamics, including their effects on land benefits. More in depth, from a methodological point of view, an exhaustive comparison between Artificial Intelligence for ES (ARIES) and Integrated Valuation of Ecosystem Services and Tradeoffs (INVEST) for modeling ES in a spatially-explicit way according to different land use scenarios in Arizona, is provided by Bagstad et al. (2013). They argue that in regions with adequate data availability, deterministic models are more appropriate (such as those included in both ARIES and INVEST), while where information are relatively scarce, probabilistic models are preferable (i.e., ARIES).

13.2 Lessons Learned from ES-Based Applications in Italy

A first attempt to evaluate the impact of urban expansion on ES in Italy has been recently released by the *Italian National Institute for Environmental Protection and Research* (ISPRA 2017) in the context of the *Soil Administration Models 4 Community Profit* (SAM4CP) project, founded by the EU LIFE + Program. From the methodological point of view, the evaluation referred to 9 ES, and the spatially-explicit evaluation of the impact of urban expansion from 2012 to 2016 countrywide, was carried out at very high resolution (10 m), using the *Copernicus High Resolution Layer* (Congedo et al. 2016) as land cover input, and the innovative dedicated models such as INVEST (Sharp et al. 2016). According to the main results, the overall yearly loss of ES due to urban expansion ranges between

625 and 907 million € per year, which correspond to 30,519/44,400 € ha^{-1}. This variability mainly depends on the variability of the economic values used to estimate each ES. In detail, in this case, 5 of the 9 ES show variability in their estimates depending on the alternative economic aspects considered in the analysis. For example, the yearly loss of carbon (C) stored ranges between 14.5 and 132.1 million € per year if considering the market value of C within a voluntary carbon market or its value expressed as a social cost for the collectivity (i.e. social cost of carbon; Sallustio et al. 2015). In light of this, we may argue that the economic evaluation of ES is strictly dependent on both the methodology and the economic values used for the evaluation, which in turn usually depend on both data availability and peculiar aspects of ES to be taken into account (i.e., market value instead of environmental and social values). For this reason, transparency and clarity of methodological aspects play a primary role for the effectiveness and reliability of the results as well as of the related planning or policy remarks (Hamel and Bryant 2017; Ruckelshaus et al. 2015).

Other findings related to the ISPRA (2017) results regard the relative contribution of the individual ES respect to the overall economic loss. It is interesting to note that crop production represents almost 54% of the total economic loss (412 million €), followed by water purification (17%), sediment retention (14%) and carbon sequestration (10%). The high contribution of the crop production ES is clearly justified and understandable if considering that almost 75% of the overall urban expansion in Italy affects agricultural land uses (ISPRA 2017; Pagliarella et al. 2016). Contrarily, for example, a relatively-low contribution of habitat quality to the overall economic loss (almost 1%) is mainly due to the low relative incidence of urban expansion on natural and seminatural land uses (22.8%; ISPRA 2017), which intuitively contribute to this ES supply. With this regard, Sallustio et al. (2015) show how the ES evaluation clearly helps to understand the relative impact of urban expansion in relation to the land use classes which are converted into urban. Specifically, in this case, even being the overall impact of urban expansion on carbon storage and sequestration higher in the Province of Rome, the higher relative impact in terms of ES was found in Molise region (12,920 and 14,186 € ha^{-1}, respectively). Authors suggest that this difference was mainly due to the different land use classes affected by land consumption in the two study areas. Hence, the higher incidence of land consumption on natural and seminatural land uses, which have a high carbon storage potentiality, rather than agricultural ones, in the Molise region, justified the higher economic loss. Trying to generalize these findings, authors conclude that «*there is a sort of anthropogenic gradient affecting the urbanization impact on C loss. Therefore, the higher the original naturalness of the territory, the higher the C loss*» (Sallustio et al. 2015, p. 85). Even if not using an economic approach, a similar finding has been recently confirmed also by Sallustio et al. (2017a), with specific regard to habitat quality and degradation in Italy.

Similarly, Rivieccio et al. (2017) found that in Abruzzo region, central Italy, the relative impact of urban expansion on crop production is directly proportional to the productivity and economic values of the croplands converted into settlements. Even if apparently obvious, if analyzed at regional scale and together with the other LUC occurred in the same time-span, this finding highlights (*i*) the relation existing among different LUC and their cumulative impacts on the overall ES variation, (*ii*) the primary role of irreversible transitions (i.e., urban expansion) to strongly undermine future ES provisioning and, more generally, sustainability at regional scale, (*iii*) the importance of the spatially-explicit analysis of ES and of its integration with others environmental, social and economic features (e.g., soil characteristics and land capability) to increase awareness of policy-makers on environmental issues such as urban expansion, and on the need to find innovative paths, i.e. the Integrated Landscape Approaches (Reed et al. 2015) to promote sustainable land use management strategies. Specifically, in this case, the economic evaluation of the impact of LUC on the agricultural sector has demonstrated that the positive effects of land use intensification processes (i.e., the shifting from low-profitable crop productions to more specialized and profitable ones, namely vineyards and olive groves) is not able to compensate the negative impact largely attributable to urban expansion in the most productive and profitable areas and land abandonment in the hilly and mountainous ones. This imbalance has a twofold effect in the medium-long term, as follows: (*i*) the overall depletion of the agricultural productivity, and related incomes, at regional scale, and (*ii*) the exacerbation of the economic discrepancy between mountain areas, affected by population ageing, outmigration and land abandonment, and lowlands, rising in population and expected to increase their productivity. This in turn will cause not only environmental, but also socio-economic impacts that will seriously jeopardize the future sustainable growth of such territories.

Some general remarks can also be provided about the methodological aspects of assessing the impacts of urban expansion on ES. At first, there is the need to develop, test and apply innovative methodologies able to maximize the informative power, accuracy and reliability of data while minimizing costs and time for their updating. This is particularly relevant in contexts where budgetary constraints, combined with the lack of standardized monitoring systems and data availability, strongly limit the possibility for policy-makers to have scientifically-sound data able to support and orient their choices on particular aspects (e.g., urban expansion in relation to its environmental impacts). In that sense, the combined use of inventory and cartographic approaches, as well as of different data sources (e.g., related to both land use and land cover aspects) have recently offered promising results at both regional and national scale in Italy (Rivieccio et al. 2017; Sallustio et al. 2015, 2016). Similarly, further efforts to optimize and maximize cost-efficiency of inventory approaches to monitor both urban expansion and other LUC (Pagliarella et al. 2016), as well as particular urban and rural landscape features such as trees outside forests or green spaces in general (Fattorini et al. 2016; Sallustio et al. 2017b), are desirable in the next future.

Finally, as recently reported by Hamel and Bryant (2017), especially when using monetary values and in applicative contexts, the assessment of the uncertainty and robustness of the results, play a key role in enhancing the credibility and trust for a variety of stakeholders and policy-makers with regard to the implementation of such tools in decision-making contexts. Obviously, the need, the robustness and the accuracy of the uncertainty analysis and in turn of the final results of the ES-analysis per se, is widely different primarily depending on the objectives of the projects (i.e., the prediction of the impacts of projects' implementation should be more rigorous, robust and accurate than exploratory, communication, or learning projects).

13.3 What's Next? Looking Beyond the Ecosystem Services-Based Approach

According to the contents above, urban expansion strongly modifies land benefits, generally leading to local economic loss and social misbalance. The ES approach incorporating trends of more impacting LUC (i.e., urban expansion) is a suitable solution to evaluate how ecosystems work, what goods and services are delivered, what is the final beneficiaries' network, and to what extent current decisions conduct to resilience. For example, Hou et al. (2015) used both biophysical and social-economic data to understand the ES flow over an urban-rural gradient. On the other hand, combining "cascade model" and *Driver-Pressure-State-Impact-Response* (DPSIR) framework may highlight the effects of human-induced environmental changes (e.g., Nassl and Löffler 2015). Through an integrated perspective, Silva et al. (2017) suggest to adopt the coupled rural-urban concept to strengthen ecosystem-based management strategies via: (*i*) assessing the territory's potentialities by ecological-economic zoning, which brings together natural assets and urban and peri-urban regions according to spatiotemporal connections; (*ii*) best distributing land production over most suitable areas across rural properties; and (*iii*) diversifying and fostering functions from rural areas (multi-functionality). In order to tackle global change issues (including effects of urbanization), the "urban-rural nexus" approach has been recently developed (i.e., Hettiarachchi and Ardakanian 2016). For example, Borgogno-Mondino et al. (2015) recently proposed a combination of landscape metrics, demographic data and soil properties to map the vocation of a peri-urban area in Italy to host new leisure facilities (i.e. green areas). Such approach represents an effective supporting tool, as it recognizes a multiple value (even economic) to soil, its properties and subsequent benefits, thus bridging the gap between land use planners (and decision-makers in general) and local stakeholders. Moreover, analyzing the complex driver-response pattern between human, land use, and ES is important to define the impact of urbanization on the surrounding environment.

The economic evaluation of the impact of urban expansion on ES clearly represents a valuable tool to inform policy-makers on the ES values otherwise often neglected during, for e.g., SEA processes of land use planning because of their economic invisibility with respect to other issues, such as urban infrastructure, settlements, etc. In this case, the ES assessment, may help to reconcile the historical bias between nature and human through ameliorating the costs-benefits analysis in decision making contexts (Jansson 2013). With this in mind, there is a rising need of improving the economic assessment of ES as well as enlarging the bundle of the considered environmental benefits (i.e., relational values; Piccolo 2017), in order to fit as much as possible to a holistic and not only to "utilitaristic" and anthropocentric, perspective.

Accordingly, on the research side, comprehensive frameworks for sustainable planning should be explored and proposed. With this perspective, Geneletti et al. (2017) have recently highlighted the importance of testing sustainable planning approaches in different contexts to reveal their potentials and limitations, as well as of assessing social-economic effects of planning approaches. Accordingly, current research should also use a social-ecological perspective. For example, adopting the landscape approaches in current research strengthens the understanding the intimate relationships between ecosystem changes and societal needs (e.g., social stratification, control of labor, and access to land; Plieninger et al. 2015).

In conclusion, impacts of urban expansion (and in general land use transformations) on land benefits may be reduced through: (*i*) fully understanding the relationship between different drivers and how they influence landowners' decision-making processes and strategies (Kristensen 2016); (*ii*) implementing nature-based solutions (NBS) to enhance the ES availability, while fostering connectivity at landscape scale (Keesstra et al. 2018); (*iii*) reducing rural marginalization by social inclusivity, promoting investments in rural sector, and reducing gaps between urban and rural economies; and (*iv*) incorporating holistic frameworks in land planning to capture economic values also from abandoned and degraded areas. Only if such strategies are simultaneously taken into account with an adaptive perspective, the negative effects of LUC on ecosystem resilience can be avoided or at least mitigated, and the social-economic gap between urban and rural contexts reduced.

References

Aretano, R., Petrosillo, I., Zaccarelli, N., Semeraro, T., & Zurlini, G. (2013). People perception of landscape change effects on ecosystem services in small Mediterranean islands: A combination of subjective and objective assessments. *Landscape and Urban Planning, 112,* 63–73. https://doi.org/10.1016/j.landurbplan.2012.12.010.

Artmann, M. (2016). Urban gray versus urban green versus soil protection—Development of a systemic solution to soil sealing management on the example of Germany. *Environmental Impact Assessment Review, 59,* 27–42. https://doi.org/10.1016/j.eiar.2016.03.004.

Bagstad, K. J., Semmens, D. J., Waage, S., & Winthrop, R. (2013). A comparative assessment of decision-support tools for ecosystem services quantification and valuation. *Ecosystem Services, 5,* 27–39. https://doi.org/10.1016/j.ecoser.2013.07.004.

Balvanera, P., Quijas, S., Karp, D. S., Ash, N., Bennett, E. M., Boumans, R., et al. (2017). Ecosystem Services. In M. Walters, R. Scholes (Eds.), *The GEO handbook on biodiversity observation networks* (pp. 39–78). Cham: Springer. https://doi.org/10.1007/978-3-319-27288-7_3.

Borgogno-Mondino, E., Fabietti, G., & Ajmone-Marsan, F. (2015). Soil quality and landscape metrics as driving factors in a multi-criteria GIS procedure for peri-urban land use planning. *Urban Forestry and Urban Greening, 14,* 743–750. https://doi.org/10.1016/j.ufug.2015.07.004
.

Brander, L. M., & Crossman, N. (2017). Economic quantification. In B. Burkhard & J. Maes (Eds.), *Mapping ecosystem services* (pp. 115–125). Sofia: Pensoft Publisher.

Burkhard, B., Crossman, N., Nedkov, S., Petz, K., & Alkemade, R. (2013). Mapping and modelling ecosystem services for science, policy and practice. *Ecosystem Services, 4,* 1–3. https://doi.org/10.1016/j.ecoser.2013.04.005.

Cardinale, B. J., Duffy, J. E., Gonzalez, A., Hooper, D. U., Perrings, C., Venail, P., et al. (2012). Biodiversity loss and its impact on humanity. *Nature, 486,* 59–67.

Congedo, L., Sallustio, L., Munafò, M., Ottaviano, M., Tonti, D., & Marchetti, M. (2016). Copernicus high-resolution layers for land cover classification in Italy. *Journal of Maps, 12*(5), 1195–1205. https://doi.org/10.1080/17445647.2016.1145151.

Costanza, R., & Daly, H. E. (1992). Natural capital and sustainable development. *Conservation Biology, 6,* 37–46. https://doi.org/10.1046/j.1523-1739.1992.610037.x.

Costanza, R., d'Arge, R., de Groot, R., Farber, S., Grasso, M., Hannon, B., et al. (1997). The value of the world's ecosystem services and natural capital. *Nature, 387,* 253–260. https://doi.org/10.1038/387253a0.

de Groot, R. S., Alkemade, R., Braat, L., Hein, L., & Willemen, L. (2010). Challenges in integrating the concept of ecosystem services and values in landscape planning, management and decision making. *Ecological Complexity, 7,* 260–272. https://doi.org/10.1016/j.ecocom.2009.10.006.

Ellis, E. C., Goldewijk, K. K., Siebert, S., Lightman, D., & Ramankutty, N. (2010). Anthropogenic transformation of the biomes, 1700 to 2000. *Global Ecology and Biogeography, 19,* 589–606. https://doi.org/10.1111/j.1466-8238.2010.00540.x.

European Commission. (2011). Communication from the commission to the european Parliament, the Council, the European Economic and Social Committee and the Committee of the Regions. *Roadmap to a Resource Efficient Europe*. COM (2011), 571 final, Brussels. http://eur-lex.europa.eu/legal-content/EN/TXT/PDF/?uri=CELEX:52011DC0571&from=EN. Accessed February 23, 2018.

European Commission. (2012). *Report from the commission to the European Parliament, the Council, the European Economic and Social Committee and the Committee of the regions: The implementation of the Soil Thematic Strategy and ongoing activities.* COM (2012), 46 final, Brussels. http://eur-lex.europa.eu/legal-content/EN/TXT/PDF/?uri=CELEX:52012DC0046&from=EN. Accessed February 23, 2018.

European Environment Agency (EEA). (2006). *Urban sprawl in Europe—The ignored challenge* (EEA report). Luxembourg: Office for Official Publications of the European Communities. https://doi.org/10.1080/02697451003740312.

European Environment Agency (EEA), Joint Research Centre (JRC). (2010). *The European Environment: State and Outlook 2010: Soil, Toxicology Letters.* https://doi.org/10.1016/j.toxlet.2010.03.051.

Fattorini, L., Puletti, N., Chirici, G., Corona, P., Gazzarri, C., Mura, M., et al. (2016). Checking the performance of point and plot sampling on aerial photo imagery of a large-scale population of trees outside forests. *Canadian Journal of Forest Research, 46*(11), 1264–1274. https://doi.org/10.1139/cjfr-2016-0013.

Fedrigotti, C., Aschonitis, V., & Fano, E. A. (2016). Effects of forest expansion and land abandonment on ecosystem services of Alpine environments: Case study in Ledro valley (Italy) for the period 1859–2011. *Global NEST Journal, 18*(4), 875–884.

Foley, J. (2005). Global consequences of land use. *Science, 309*(5734), 570–574. https://doi.org/10.1126/science.1111772.

Gardi, C. (2017). *Urban expansion, land cover and soil ecosystem services*. New York: Routledge. https://doi.org/10.4324/9781315715674.

Geneletti, D., La Rosa, D., Spyra, M., & Cortinovis, C. (2017). A review of approaches and challenges for sustainable planning in urban peripheries. *Landscape and Urban Planning, 165,* 231–243. https://doi.org/10.1016/j.landurbplan.2017.01.013.

Grêt-Regamey, A., Bebi, P., Bishop, I. D., & Schmid, W. A. (2008). Linking GIS-based models to value ecosystem services in an Alpine region. *Journal of Environmental Management, 89,* 197–208. https://doi.org/10.1016/j.jenvman.2007.05.019.

Griggs, D., Stafford-Smith, M., Gaffney, O., Rockström, J., Ohman, M. C., Shyamsundar, P., et al. (2013). Policy: Sustainable development goals for people and planet. *Nature, 495,* 305–307. https://doi.org/10.1038/495305a.

Grimm, N. B., Faeth, S. H., Golubiewski, N. E., Redman, C. L., Wu, J., Bai, X., et al. (2008). Global change and the ecology of cities. *Science, 319,* 756–760. https://doi.org/10.1126/science.1150195.

Haase, D., Kabisch, N., & Haase, A. (2013). Endless urban growth? On the mismatch of population, household and urban land area growth and its effects on the urban debate. *PLoS ONE, 8*(6), e66531. https://doi.org/10.1371/journal.pone.0066531.

Haase, D., Haase, A., & Rink, D. (2014). Conceptualizing the nexus between urban shrinkage and ecosystem services. *Landscape and Urban Planning, 132,* 159–169. https://doi.org/10.1016/j.landurbplan.2014.09.003.

Haines-Young, R., & Potschin, M. (2013). *Common international classification of ecosystem services* (CICES, Version 4.3). Report to European Environment Agency, pp. 1–19. https://doi.org/10.1016/b978-0-12-419964-4.00001-9.

Hamel, P., & Bryant, B. P. (2017). Uncertainty assessment in ecosystem services analyses: Seven challenges and practical responses. *Ecosystem Service, 24,* 1–15. https://doi.org/10.1016/j.ecoser.2016.12.008.

Hennig, E. I., Schwick, C., Soukup, T., Orlitová, E., Kienast, F., & Jaeger, J. A. G. (2015). Multi-scale analysis of urban sprawl in Europe: Towards a European de-sprawling strategy. *Land use policy, 49,* 483–498. https://doi.org/10.1016/j.landusepol.2015.08.001.

Hettiarachchi, H., & Ardakanian, R. (2016). Managing water, soil, and waste in the context of global change. In H. Hettiarachchi & R. Ardakanian (Eds.), *Environmental resource management and the nexus approach* (pp. 1–7). Cham: Springer. https://doi.org/10.1007/978-3-319-28593-1.

Hou, Y., Mller, F., Li, B., & Kroll, F. (2015). Urban-rural gradients of ecosystem services and the linkages with socioeconomics. *Landscape Online, 39,* 1–31. https://doi.org/10.3097/LO.201539.

Italian National Institute for Environmental Protection and Research (ISPRA). (2016). *Consumo di suolo, dinamiche territoriali e servizi ecosistemici.* Rome.

Italian National Institute for Environmental Protection and Research (ISPRA). (2017). *Consumo di suolo, dinamiche territoriali e servizi ecosistemici.* Rome.

Jansson, Å. (2013). Reaching for a sustainable, resilient urban future using the lens of ecosystem services. *Ecological Economics, 86,* 285–291. https://doi.org/10.1016/j.ecolecon.2012.06.013.

Kabisch, N., & Haase, D. (2011). Diversifying European agglomerations: Evidence of urban population trends for the 21st century. *Population Space Place, 17,* 236–253. https://doi.org/10.1002/psp.600.

Keesstra, S., Nunes, J., Novara, A., Finger, D., Avelar, D., Kalantari, Z., et al. (2018). The superior effect of nature based solutions in land management for enhancing ecosystem services. *Science of the Total Environment, 610–611,* 997–1009. https://doi.org/10.1016/j.scitotenv.2017.08.077.

Kristensen, S. B. P. (2016). Agriculture and landscape interaction landowners' decision-making and drivers of land use change in rural Europe. *Land Use Policy, 57,* 759–763. https://doi.org/10.1016/j.landusepol.2016.05.025.

Kroll, F., Müller, F., Haase, D., & Fohrer, N. (2012). Rural-urban gradient analysis of ecosystem services supply and demand dynamics. *Land Use Policy, 29,* 521–535. https://doi.org/10.1016/j.landusepol.2011.07.008.

Layke, C., Mapendembe, A., Brown, C., Walpole, M., & Winn, J. (2012). Indicators from the global and sub-global millennium ecosystem assessments: An analysis and next steps. *Ecological Indicators, 17,* 77–87. https://doi.org/10.1016/j.ecolind.2011.04.025.

Li, B., Chen, D., Wu, S., Zhou, S., Wang, T., & Chen, H. (2016). Spatio-temporal assessment of urbanization impacts on ecosystem services: Case study of Nanjing City, China. *Ecological Indicators, 71,* 416–427. https://doi.org/10.1016/j.ecolind.2016.07.017.

Mace, G. M., Norris, K., & Fitter, A. H. (2012). Biodiversity and ecosystem services: a multilayered relationship. *Trends in Ecology and Evolution, 27,* 19–26. https://doi.org/10.1016/j.tree.2011.08.006.

MEA. (2005). *Millennium Ecosystem Assessment. Ecosystems and human well-being: Synthesis.* Washington DC: Island Press.

Mendoza-González, G., Martínez, M. L., Lithgow, D., Pérez-Maqueo, O., & Simonin, P. (2012). Land use change and its effects on the value of ecosystem services along the coast of the Gulf of Mexico. *Ecological Economics, 82,* 23–32. https://doi.org/10.1016/j.ecolecon.2012.07.018.

Meyfroidt, P., & Lambin, E. F. (2011). Global forest transition: Prospects for an end to deforestation. *Annual review of environment and resource, 36,* 343–371. https://doi.org/10.1146/annurev-environ-090710-143732.

Nassl, M., & Löffler, J. (2015). Ecosystem services in coupled social–ecological systems: Closing the cycle of service provision and societal feedback. *Ambio, 44,* 737–749. https://doi.org/10.1007/s13280-015-0651-y.

Pagliarella, M. C., Sallustio, L., Capobianco, G., Conte, E., Corona, P., Fattorini, L., et al. (2016). From one- to two-phase sampling to reduce costs of remote sensing-based estimation of land-cover and land-use proportions and their changes. *Remote Sensing of Environment, 184,* 410–417. https://doi.org/10.1016/j.rse.2016.07.027.

Paleari, S. (2017). Is the European Union protecting soil? A critical analysis of Community environmental policy and law. *Land use policy, 64,* 163–173. https://doi.org/10.1016/j.landusepol.2017.02.007.

Palomo, I., Bagstad, K. J., Nedkov, S., Klug, H., Adamescu, M., & Cazacu, C. (2017). Tools for mapping ecosystem services. In B. Burkhard & J. Maes (Eds.), *Mapping ecosystem services* (pp. 72–76). Sofia: Pensoft Publisher.

Piccolo, J. J. (2017). Intrinsic values in nature: Objective good or simply half of an unhelpful dichotomy? *Journal for Nature Conservation, 37,* 8–11. https://doi.org/10.1016/j.jnc.2017.02.007.

Plieninger, T., Kizos, T., Bieling, C., Dû-Blayo, L. Le, Budniok, M. A., Bürgi, M., et al. (2015). Exploring ecosystem-change and society through a landscape lens: Recent progress in european landscape research. *Ecology and Society, 20,* 1–10. https://doi.org/10.5751/ES-07443-200205.

Reed, J., Deakin, L., & Sunderland, T. (2015). What are "Integrated Landscape Approaches" and how effectively have they been implemented in the tropics: A systematic map protocol. *Enviromental Evidence, 4,* 1–7. https://doi.org/10.1186/2047-2382-4-2.

Rivieccio, R., Sallustio, L., Paolanti, M., Vizzarri, M., & Marchetti, M. (2017). Where land use changes occur: Using soil features to understand the economic trends in agricultural lands. *Sustainability, 9,* 1–20. https://doi.org/10.3390/su9010078.

Rodríguez, J. P., Beard, T. D., Jr., Bennett, E. M., Cumming, G. S., Cork, S. J., Agard, J., et al. (2006). Trade-offs across space, time, and ecosystem services. *Ecology and society, 11,* 1–14. https://doi.org/10.5751/ES-01667-110128.

Romano, B., Zullo, F., Fiorini, L., Ciabò, S., & Marucci, A. (2017). Sprinkling: An approach to describe urbanization dynamics in Italy. *Sustainability, 9,* 1–17. https://doi.org/10.3390/su9010097.

Ruckelshaus, M., McKenzie, E., Tallis, H., Guerry, A., Daily, G., Kareiva, P., et al. (2015). Notes from the field: Lessons learned from using ecosystem service approaches to inform real-world decisions. *Ecological Economics, 115,* 11–21. https://doi.org/10.1016/j.ecolecon.2013.07.009.

Runting, R. K., Bryan, B. A., Dee, L. E., Maseyk, F. J. F., Mandle, L., Hamel, P., et al. (2017). Incorporating climate change into ecosystem service assessments and decisions: A review. *Global Change Biology, 23,* 28–41. https://doi.org/10.1111/gcb.13457.

Sallustio, L., Quatrini, V., Geneletti, D., Corona, P., & Marchetti, M. (2015). Assessing land take by urban development and its impact on carbon storage: Findings from two case studies in Italy. *Environmental Impact Assessment Review, 54,* 80–90. https://doi.org/10.1016/j.eiar.2015.05.006.

Sallustio, L., Munafò, M., Riitano, N., Lasserre, B., Fattorini, L., & Marchetti, M. (2016). Integration of land use and land cover inventories for landscape management and planning in Italy. *Environmental Monitoring and Assessment, 188,* 1–20. https://doi.org/10.1007/s10661-015-5056-7.

Sallustio, L., De Toni, A., Strollo, A., Di Febbraro, M., Gissi, E., Vizzarri, M., et al. (2017a). Assessing habitat quality in relation to the spatial distribution of protected areas in Italy. *Journal of Environmental Management, 201,* 129–137. https://doi.org/10.1016/j.jenvman.2017.06.031.

Sallustio, L., Perone, A., Vizzarri, M., Corona, P., Fares, S., Cocozza, C., et al. (2017b). The green side of the grey: Assessing greenspaces in built-up areas of Italy. *Urban Forestry and Urban Greening*. https://doi.org/10.1016/j.ufug.2017.10.018. (in press).

Salvati, L., Munafo, M., Morelli, V. G., & Sabbi, A. (2012). Low-density settlements and land use changes in a Mediterranean urban region. *Landscape and Urban Planning, 105,* 43–52. https://doi.org/10.1016/j.landurbplan.2011.11.020.

Sharp, R., Tallis, H.T., Ricketts, T., Guerry, A.D., Wood, S.A., Chaplin-Kramer, R., et al. (2016). *InVEST 3.3.1 user's Guide*. Palo Alto, CA, USA: The Natural Capital Project.

Silva, R. F. B., da Rodrigues, M. D. A., Vieira, S. A., Batistella, M., & Farinaci, J. (2017). Perspectives for environmental conservation and ecosystem services on coupled rural–urban systems. *Perspectives in Ecology and Conservation, 15*(2), 74–81. https://doi.org/10.1016/j.pecon.2017.05.005.

Steffen, W., Richardson, K., Rockström, J., Cornell, S., Fetzer, I., Bennett, E., et al. (2015). Planetary boundaries: Guiding human development on a changing planet. *Science, 347*(6223), 736–747. https://doi.org/10.1126/science.aaa9629.

Stolte, J., Tesfai, M., & Keizer, J. (2016). Soil threats in Europe: Status, methods, drivers and effects on ecosystem services (JRC Technical Reports). Luxembourg: Soil Threats in Europe.

TEEB. (2010). *The economics of ecosystems and biodiversity: Mainstreaming the economics of nature: A synthesis of the approach, conclusions and recommendations of TEEB*.

United Nations. (1992). Convention on biological diversity. In *Diversity* (p. 30). https://doi.org/10.1146/annurev.ento.48.091801.112645.

Zhang, Y., Zhao, L., Liu, J., Liu, Y., & Li, C. (2015). The impact of land cover change on ecosystem service values in urban agglomerations along the coast of the Bohai Rim, China. *Sustainability, 7,* 10365–10387. https://doi.org/10.3390/su70810365.

Marco Marchetti is Full Professor (Chair of Forest Management and Conservation) and Vice-Rector at the University of Studies of Molise (Italy). Since 2012 he is Chairman of the Scientific Advisory Board of the European Forest Institute, and since 2015 of the Italian Society of Forest Ecology and Silviculture (SISEF) and of the Research Centre for Inner Areas and Apennines (ArIA), and, from 2017, Chairman of the Italian Association of Agricultural Science Societies (AISSA). Current research mainly focuses on multiresources inventory and land use mapping, forest and land use management and planning with particular regard to Natural Capital and protected areas.

Matteo Vizzarri is a research fellow. He is mostly interested in forest ecosystem modeling, particularly simulating the effects of forest management and planning on goods and services' provision. His research activity currently relates with the fields of bio-economy and sustainable resources use, especially through developing topics about for example, conservation of key species and habitats, climate change adaptation and mitigation, wood and en-ergy chains, and participatory decision making.

Lorenzo Sallustio is a fixed-term researcher. His research activity is principally focused on land use and land use change monitoring at different spatial scales (from regional to national) and the assessment, mapping and evaluation (from both biophysical and economic perspectives) of their impacts on ecosystem services (principally biodiversity, carbon storage and sequestration, and timber provisioning). He is primarily interested on land abandonment, forest recolonization and urban growth dynamics. Currently, he is also involved in developing innovative sustainable strategies for rural areas through valorizing their Natural Capital in the framework of the Italian Strategy for Inner Areas.

Chapter 14
Ecosystems Services and Spatial Planning: Lessons Learned from the Life SAM4CP Project

Carolina Giaimo, Carlo Alberto Barbieri and Stefano Salata

14.1 Land Consumption and Spatial Planning

For over a decade—though more recently in Italy—international attention has been focussed on the theme of limiting land consumption. This attention has gradually spread out further and focalised (i) on the matter of this vital objective—for the ecosystemic, landscaping, productive and socio-economic values to be assigned to land insofar as it is a fundamental finite resource, (ii) as well as for the legislative-procedural aspect related to the way to plan the territory to address this issue (especially in Italy).

Regarding the latter aspect, the common view should be that of the INU (the *Italian National Town Planning Institute*) which believes that the best way to limit land consumption is not merely as a result of sectorial legislation but that the most suitable decision-making and evaluation ambit is in fact in territorial management over the various scales[1]; furthermore, this aspect should be innovated and relaunched as a fundamental activity of territorial management (Barbieri 2017), interpreting and evaluating environmental and settlement conditions and vocations,

[1]In the sense of obliging Regional territorial and landscape planning, the structural planning of metropolitan cities, that of Province coordination and above all the Urban Plans of the Union of smaller municipalities to express, through structural interpretations of the territory, "invariant" and precise rules and regulations aimed at avoiding the transformation of non-urbanised territories unless found necessary following careful evaluation of the alternative possible reuse of abandoned or under-used areas or the possibility of renewal of the areas and the necessary parameters of ecological sustainability and limitation, mitigation and compensation.

C. Giaimo (✉) · C. A. Barbieri · S. Salata
DIST, Interuniversity Department of Regional and Urban Studies and Planning, Politecnico and Università Di Torino, Viale Mattioli 39, 10125 Turin, Italy
e-mail: carolina.giaimo@polito.it

E. Gottero (ed.), *Agrourbanism*, GeoJournal Library 124,
https://doi.org/10.1007/978-3-319-95576-6_14

forecasts and planning through effective *Strategic Environmental Assessment* (SEA). Moreover, this activity must make the most of the co-planning[2] method in which it is more feasible—within an ambit of governance and sustainable execution of the principle of institutional subsidiarity—to procedurally share the choices and opportunities but also the constraints and limitations that the traditional hierarchical and divided planning system has never been able to express or bring into effect substantially.

14.1.1 The Project LIFE + Soil Administration Model 4 Community Profit

The themes that LIFE SAM4CP[3] deal with are (i) land as a non-renewable resource and the supply of its fundamental *Ecosystem Services* (ESs) considered as common property and values, (ii) the consequently necessary limitation of land consumption and constructive use of the same through conscientious territorial management starting on the town planning scale.

In order to achieve the above objectives, the project has a twofold ideal:

- the importance of spatial planning (in particular in the Metropolitan City), especially in municipalities or the Union of the municipalities (which is conformative to the property of the land and regulation of use thereof), as activities aimed at guaranteeing overall sustainability and savings (also in public funding) to communities by preserving the ecosystemic resources of the land;
- the importance of conscientiously taking on the biophysical and economic value of local Ecosystem Services in choices regarding territorial management, especially in decision-making processes regarding spatial planning and local communities. Within this perspective, constructive support for town planning decisions can be found in those activities of the Project that produce simulation models (scenarios) on the impacts/effects of predicted consumption of land; models to which we can associate the qualitative quantification of the biophysical and economic effects of various planning choices.

More generally speaking, the new role of resilience planning requires that we work on the goals, new paradigms and contents of spatial planning at local level. This refers in particular to the theme of *urban regeneration* (or rather, *regeneration of the territory* if we refer to vast settlements and areas of *sprawl*) linked to the aim of limiting land take and valorising its Ecosystem Services. The city and territory

[2]Refers to the procedure conceived and regulated by Regional Legislation of Piedmont with Regional Laws 1/2007 and 3/2013.

[3]The Project (see: http://www.sam4cp.eu/en/) has been coordinated by the Metropolitan City of Turin with the Interuniversity Department of Regional and Urban Studies and Planning of the Polytechnic and the University of Turin (DIST), ISPRA, CREA and CSI Piemonte.

project should focus on the possible reuse of the urbanised territory and the integrated physical, socio-economic and ecologic regeneration of the urban welfare thereof (and not only of the more obsolete parts of the city) and the settlement and infrastructural systems.

This stage signals the complete abolition of spatial planning based on expansion—which has been a long time to come and can no longer be put off—in favour of a more innovative discipline of flexible and thus resilient planning. That requires an equally necessary reform in town planning and the technical and regulatory instruments thereof.

14.1.2 The Government of Sustainable Changes of Land Use

The territorial government is a broad concept composed both by technical and political skill and expertise that we can't address to a "discipline" in the traditional sense: hence it can't be limited to the regulation of land use and building. It represents an integrated system of theories and practices made by knowledge and experiences with a proper horizontal and multiscalar "functional role", which exceeds the ordinary skills fragmentation of urban planning disciplines. Indeed, the territorial government includes a vast amount of expertise: environment, landscape, soil conservation, ecosystems protection, enhancement of cultural, historical and environmental heritage, socio-economic development, mobility and territorial infrastructure.

Adopting rules and procedures for innovative urban and territorial planning is essential to support policies for the protection, enhancement and qualification of the settlements, for urban regeneration and reduction of land take as well as governing climate change. But innovations for sustainability requires to assume the principle of subsidiary and to recognize that territorial government is an integrated "horizontal" process rather than a "hierarchic system" composed by separated plans drawn up by different and separated institutions.

By the way, there is a widespread awareness that the Italian hierarchical planning model (and related procedures) based on the approval through dirigisme is now obsolete and inadequate (Barbieri and Giaimo 2015). Nowadays the environmental question re-emerges with a different, more dramatic impact, in the measure in which contemporary cities and territories are exposed to the dangers caused by climate change and its effects on urban life, combined with other physical risks (seismic activity, damaging hydrogeological events, air pollution) and, above all, those linked to the global economic crisis and that of public finances, the emergence of new inequalities, poverty and social tension. Therefore, the sustainability challenge is a matter of institutional relations, rather than a matter of technology for planning: relationships must become more horizontal and based at all on methods and procedures for cooperation between local authorities and consultation/ participation of public and private stakeholders.

14.1.2.1 Sharing Knowledge and Building Consensus

The LIFE SAM4CP project is based on the conviction that innovative transition from urban planning to local governance is intimately connected with a multilevel system of shared knowledge and methods. When decision making and public deliberation processes are characterized by a multiplicity of public subjects with relevant and differentiated tasks, it is essential to practice the multilevel governance.

It is of particular significance in the field of urban and territorial transformations because it deals with the uniqueness of the physical space and the natural interdependence between the various components of the environment (air, water, topsoil, subsoil, biotic communities, etc.) which characterize any human activity and conditions for its operability. Such environmental organic unity and integration contradicts the separation and segmentation of tasks and functions that describe the administrative action.

Practicing multilevel governance means to implement actions, behaviours and attitudes that favour a process of decision-making avoiding the deliberations of authority, which in turns, imply decision makers, primarily public, may, as a consequence, don't adopt attitudes that determine the block of each operability.

Considering that in the Italian public administration persist a hierarchical, top-down approach, where public subjects express their own formal "veto power" to "influence" local policy, it is, therefore, necessary that public administrations, at all levels—and private operators too—undertake attitudes to co-work for achieving sustainability. The LIFE SAM4CP assume that a good solution consists in practising the co-planning attitude for the territorial government, through the instrument of the Co-planning Conference.

In Piedmont Region, the introduction of this new approach to urban planning dates back to the *Regional Law n.1 of 2007* which was a partial modification of the regional planning law in force at the time. Then it was confirmed by further partial amendment introduced by the *Regional Law n. 3 of 2013*.

Among the activities of co-planning, the relationships between public institutions and between institutions and users, are based on collaboration and participation in the definition of planning contents. In that phase, the involved subjects are forced to share the description of the knowledge framework (which is complicated and multileveled among different scales) and the objectives, methods and projects.

Therefore LIFE SAM4CP experiments a co-planning activity where every institution provides their information, knowledge, skills and specificities, in particular through its plans, and the local administration can share its expertise and experience and compares its ecosystem services-based evaluations with other institutions and stakeholders, seeking incrementally an agreement on the general objectives and actions to pursue.

14.2 Ecosystem Services Paradigm for Integrating Spatial Planning and Environmental Assessment

Integrating skills—at all scales—of ecological, economic and socio-political disciplines within a transdisciplinary framework is the challenge of LIFE SAM4CP project, so as ES assessment for planning purposes is one of the challenges that both academic and administrative sectors have to deal in the next years. The ES analysis is one that requires such integration between different evaluations, because the context based assessment of biophysical and economic values is based on integration between theories of environmental economy, geographic information system, mapping and representation of spatial data.

The role of the ES analysis should enforce the integrated planning approach, especially joining the SEA that produces planning scenarios and a shared framework for evaluates, by the public and private subjects operating at different levels and in different sectors, the spatial trade-offs among different land use functions. The role of ES assessment is to define the fixed and flexible elements of negotiations for land use regulation. Therefore, it represents the integration between plan and SEA which is essential to identify broad strategies of sustainable development overcoming the technological enhancement of environmental issues.

The ES analysis helps institution and local stakeholders to play their choices within a strategy for sustainable development, which fixes the rules for improving or to restore the characters of the ecological functionality of the territory. However, this methodology doesn't have adequate regulatory support in the planning process.

For this reason, one of the LIFE SAM4CP actions foresees the use of mapping and evaluation of ESs as a starting point for drawing a structural variant of the *Local Land Use Plan* in four selected municipalities of the *Metropolitan City of Turin* (the selection comprises Bruino as a pilot case, Chieri, Settimo Torinese and None). These municipalities, whose land use variants pursue the goal of reducing land consumption, take part in the Co-planning Conference with the analysis on ES developed during the project. Such analytical framework is generated before the final definition of the contents of the plan and at the same time with the SEA.

14.2.1 Methodology

The project was aimed at producing a spatial prefiguration of ES assessment within a multidisciplinary perspective and downscaling the spatial evaluation from the National to the Local scale (Maes et al. 2012; Pulighe et al. 2016). The main challenge posed by the project was the support of decision-making processes at the local level aimed at designing land use planning tools. Taking on this goal as the main target of the research, the project was the first to overcome the traditional separation between the theoretical knowledge of ES and the practical application thereof in planning (Maes et al. 2012; Kaczorowska et al. 2016; Langemeyer et al. 2016).

The ES maps cover three out of the four ES categories according to *The Economics of Ecosystems & Biodiversity* (TEEB) program (Partidario and Gomes 2013):

1. Habitat Quality (supporting),
2. Carbon Sequestration (regulating),
3. Water Yield (regulating),
4. Nutrient Retention (regulating),
5. Sediment Retention (regulating),
6. Crop Production (provisioning),
7. Crop Pollination (provisioning).

Unfortunately, the cultural category was overlooked by the project as finding adequate indicators for mapping proved rather difficult during the early phases of research (the project started in 2014). Nonetheless, the "regulative" component covers the majority of ESs, thus it was possible to address a multifunctional perspective to define a composite indicator, namely the *Ecosystem Service Value* (ESV) (Clerici et al. 2014; Salata and Gardi 2014; Alam et al. 2016) which has been used to assess how each category of Land Use behaves in respect to ES provisions.

The mapping activity in the cases study was completed using the *Integrated Valuation of Ecosystem Services and Tradeoffs software* (InVEST version 3.2.0) (Tallis et al. 2011). This software is highly useful in addressing sustainability over decision-making processes when strategies and quantitative targets are defined by stakeholders involved in designing the land use plans (Mubareka et al. 2011; Grêt-Regamey et al. 2017). Current and scenario analysis under different land use configurations is the core of the ES assessment as defined by the project methodology.

To fulfil the goal, every ES map has been designed to have a high level of accuracy and precision for planning purposes: the challenge was not to use InVEST as a general tool for ES accountability but rather to evaluate how prescriptive land use scenarios should be compared during the Co-Planning phase (Salata et al. 2017).

A further part of the work was dedicated to running and testing the InVEST software for each ES selected. In particular, actions were dedicated to:

- the construction dataset (using standard and ancillary data);
- the research of sources for input software values;
- the interpretation of output models.

Biophysical model outputs were distributed over five-meter raster cells and, subsequently, economic values were associated thereto to estimate the value of each ES (Laurans et al. 2013; Silvennoinen et al. 2017). The biophysical evaluation produces output per pixel expressed by (i) indexes or (ii) absolute quantities. The seven ESs mapped by the project were estimated using such units (see Table 14.1):

Table 14.1 Ecosystem Services and their values

Ecosystem function	Biophysical value	Economic value
Habitat quality	The overall quality of the ecosystem (biodiversity) [index 0–1]	Cost of the "reproduction" of specific land uses that provides ES [20 €/sq.m.]
Carbon sequestration	Tons of sequestered carbon by soil [t/px]	Price for each ton of carbon stored [120 €/mc]
Water yield	Liters of water removed by processes of evapotranspiration [mm/px]	Cost for removing water by artificial techniques as a construction of a lamination hydro-basin [12.6 €/mc]
Nutrient retention	Nitrates released into the water [kg/px]	Cost for the construction of green buffer zones useful to detention of nitrates [64 €/kg]
Sediment retention	Potential erosion avoided by soil [t/px]	Cost of rehabilitation of soil fertility. useful to the protection from erosion [22.8 €/t]
Crop pollination	Gradient of optimal allocation for hives [index 0–1]	Average price of hive [44 €/hive]
Crop production	Standard production [€/ha.]	Prices of specific crops [€/ha.]

Source LIFE SAM4CP—Action B5

- index from 0 to 1 for Habitat Quality and Crop Pollination;
- tons/pixel for Carbon Sequestration and Sediment Retention; mm/pixel for Water Yield; kg/pixel for Nutrient Retention;
- euro/pixel for Crop Production.

The model outputs are briefly described below:

- *Habitat Quality*. The map shows the cluster where the quality of non-specific habitats (as proxy of the overall environmental quality) is high or low. The map recognizes the locations of the main patches of natural values. In the middle of the flat valley-floor where the settlement system is distributed, a leapfrogged morphology of medium environmental quality patches are even found close to dense residential and industrial zones.
- *Carbon Sequestration*. The map shows which soils are capable of "holding" a high degree of carbon and provide a "carbon pool" function which is also fundamental for the climate change mitigation policy.
- *Water Yield*. The map represents the relative contributions of each land use cell in retaining water per watershed. The value of evapotranspiration has been used to map those areas that best retain water from the stream via evapotranspiration.
- *Nutrient Retention*. The map indicates the contribution of vegetation and soil in purifying water through the removal of nutrient pollutants from runoff.

In particular, the map shows at pixel level how much load from each pixel eventually reaches the stream.

- *Sediment Retention*. The map indicates the total potential soil loss per pixel in the original land cover. High values correspond to places where loss of sediment is higher than in other parts. This vulnerability is crucial especially in territories of high risk of landslides (hills or mountains).
- *Crop Pollination*. The map represents an index of the likely abundance of pollinator species on each agricultural cell in the landscape. It provides information on agricultural sites that have better provision benefitting from natural pollination;
- *Crop Production*. This service has been mapped using the Standard Production of the different agricultural cell and not InVEST.

14.2.1.1 The Context of Study

As mentioned above, the mapping activity has been tested against the study area comprising four municipalities of the Metropolitan city of Turin. For this purpose, different kinds of municipalities were selected during the initial phase of the research according to the need to cover a wide spectrum of the morphological conditions of the metropolitan territory (see Fig. 14.1).

The case of Bruino represents a small test territory of the second-ring municipalities and the quick feedback supplied by the ES evaluation of Land Use changes in various scenarios was very useful. Then three other cases where selected representing relevant examples in the metropolitan context and belonging also to different socio-morphological characterizations: Settimo Torinese is a first-ring city in the most highly urbanized and infrastructured axes connecting Turin with Milan; Chieri is a second-ring city in the hilly and rural part of the metropolitan context with a very scenic landscape and a relevant natural environment and, finally; None is a second-ring city in the plain land and low valley area which belongs to the rural-metropolitan territory subject to considerable industrial expansion and which now is suffering the economic crisis trying to reconvert its abandoned built-up stock. Overall, their urban planning conditions were significantly different regarding the needs and character of the community, the local economy, and the morphological state.

In particular (see Table 14.2), Bruino spans 566 ha, is made up thus: 60% anthropic areas (including urban green belts) and 34% agricultural land. Its dimension was of great help to check if a minimum Land Use variation was registered by ES indicators (e.g., the sensibility of ES indicators to Land use change); and it provided an excellent feedback in adjusting and defining the model calibration.

Settimo Torinese spans 3.328 ha and is made up thus: 55% anthropic areas (including urban green belts) and less than 42% agricultural land. It is a hard industrial peripheral area with low landscape and environmental value due to the

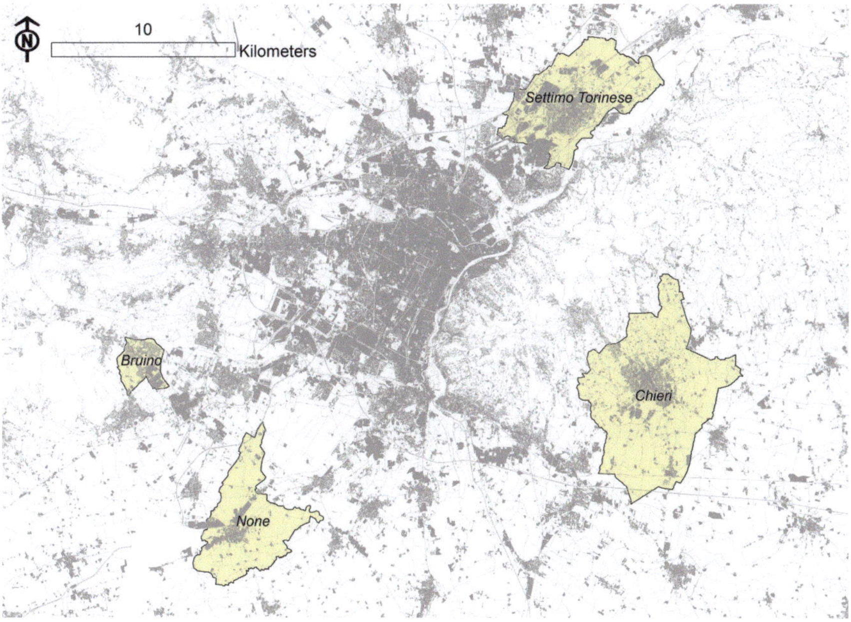

Fig. 14.1 The context of study. *Source* LIFE SAM4CP—Actions B5 and B6. Cartography: Urban areas- BDTRE Regione Piemonte

Table 14.2 Land use and land cover composition in the four cases of study

	Chieri		Bruino		None		Settimo	
LULC	abs	(%)	abs	(%)	abs	(%)	abs	(%)
Urban areas	995.22	18.37	297.92	52.55	374.57	15.20	1499.40	45.05
Green urban areas	92.89	1.72	46.07	8.13	33.84	1.37	318.83	9.58
Agricultural land	3995.96	73.77	194.64	34.33	1974.33	80.12	1384.32	41.59
Natural and seminatural	332.40	6.14	28.30	4.99	81.33	3.30	126.09	3.79
Total	5416.47	100.00	566.94	100.00	2464.08	100.00	3328.64	100.00

Source LIFE SAM4CP Actions B5 and B6

fragmentation of its residual rural system. The Administration is nowadays struggling to reconvert significant brownfields and lower the build-up capacity under the current economic crisis.

None spans 2.464 ha and is made up thus: 21% anthropic areas (including urban green belts) and more than 80% agricultural areas with a high rural characterization, nonetheless including huge abandoned areas.

Chieri spans 5.416 ha, less than 20% of its territory is covered by anthropic areas (including urban green belts) and 73% of the area is set aside for agricultural uses. It is the only case in which the natural areas display a value over 6%. Thus, this rural territory is characterized by a strong presence of natural or agroforestry uses. The Administration was involved in policy cutting back on building rights for industrial purposes that in the last 20 years had been confirmed by land use plans that inflated the capacity thereof.

Overall, the context of study covers an area spanning over 11,776 ha, meaning that 0.46% of the metropolitan territory is covered by the ESV data that highly represents the different socio-morphological character of the Metropolitan City.

To overcome the limitation of a single ES evaluation in the study area, the project takes on a multidisciplinary approach in defining where and how multi-functional ES provisioning is addressed thus reaching a quick-and-easy representation of the overall quality of the soil. This approach was undertaken in order to ease assessment during decision-making processes for land use planning (Antrop 2004; Kaczorowska et al. 2016; Salata 2017). Moreover, the visualization of the ES soil values within a single indicator proved useful in understanding how the overall ES provisioning is distributed among the different Land Use classes and mainly where the agro-rural system can provide various ES performances.

Describing the performance of agricultural areas in terms of ES provided by the soil is a complex operation. Agriculture, in fact, uses the services provided by natural capital and in turn influences them, for example by increasing the infiltration of nutrients contained in fertilizers such as nitrogen and phosphorus in the soil, but at the same time provides an essential service as a primary source of food. The land take process completely cancels the Crop production ES by making it impossible to exploit the soil of agricultural land in the short or medium-long term; in fact, the soil is a resource that requires long periods to recover its functions and its original biological characteristics. Therefore in all artificial areas, the value of this service is zero.

One result of the research was a comparison of the ESV performance between the agro-rural system and the one supplied by the anthropic landscape.

One of the significant assumptions of soil quality evaluation is that land use changes—and particularly the process of anthropization affecting rural areas—completely neglect the possibility of delivering other ES typologies in the future. Often such trade-offs between urban and rural uses oversimplify the complex relationships between many ES and the different kinds of anthropic uses that urban and suburban built-up environments provide. For example, it has been demonstrated how suburban agro-rural areas should deliver more ES than rural areas in the countryside, just as urban green areas are often more beneficial to the community that those located further from the centre and in areas of higher density (Setälä et al. 2014; Schröter et al. 2015; Turkelboom et al. forthcoming). Moreover, ESV has been used to create a cross tabulation analysis of Land Uses concerning the four different municipalities selected. Results demonstrate autonomous patterns, and an initial discussion will be presented hereafter.

14.2.2 Results and Discussion: The Distribution of the Ecosystem Service Value

The ESV indicator was employed using the Esri ArcGis (Ver. 10.5) weighted overlay function. It accounts for five out of the seven ESs selected by the study. The ESV is the sum of the Habitat Quality, Carbon Sequestration, Water Yield, Nutrient Retention and Sediment Retention values. All five indicators were normalized, meaning the final composite score ranged from 0 (worst performance, e.g., a pixel that does not deliver any ES) to 5 (best performance, all pixels offer the maximum ES capacity). Subsequently, the total score was re-classified from 0 to 1 instead of 0 to 5 for better comprehension. The ESV scores assigned to the table were the average per pixel value of the ESV in the selected Land Use.

The Table 14.3 shows how ESV performs in the aggregated Land Uses in the SAM4CP municipalities. None obtained best composite performance; the best performing was Settimo Torinese followed by Chieri and Bruino in joint second ranking. This complex ESV value is a relative count derived from the average of the individual performances. Surprisingly, Settimo Torinese performed better than Chieri and Bruino in the composite value despite threats posed by anthropic pressures with higher intensity. An in-depth assessment of the individual Land Use performances reveals that the above ranking can be explained by Bruino's negative performance specifically in ES Carbon Sequestration (0.167, Bruino has the smallest territorial surface) and Chieri's ranking last specifically in ES Sediment Retention (0.365, Chieri is located in hilly territory and is thus affected by higher erosion).

Considering only Habitat Quality as a proxy of the overall ecological value, Chieri ranks top as expected (0.400), followed by None (0.383), Bruino (0.370) and Settimo Torinese (0.345).

Concerning ESV performance in Land Use class specifically, the best value for the agricultural use is performed by None (0.362), then follows Bruino, Settimo Torinese and Chieri (respectively 0.319, 0.291 and 0.287).

The best ESV performance in urban areas was represented by None (0.254), while Bruino, Settimo Torinese and Cheri follow with lower values. This ranking refers to the capability of dense urban areas to deliver multiple ESs. The urban morphology of None and Bruino is characterized by detached or semi-detached houses with an abundance of private gardens guaranteeing excellent ecosystemic conditions; while in Chieri and Settimo Torinese, despite the significant presence of public green areas, the overall ecosystemic quality remains rather low. This situation goes far in explaining that in order to deliver multiple ecosystems in urban areas, the "porosity" index (rather than the sealing degree) plays an independently crucial role regarding property rights (public or private green areas).

According with Butsic et al. (2017), we have to be aware that the results depend on the set of ESs selected: if other ESs where chosen, the outcome might have been different.

Table 14.3 Ecosystem service value assessment: models results

Chieri	Habitat quality	Carbon sequestration	Water yield	Nutrient retention	Sediment retention	(avg)
Urban areas	0.056	0.079	0.067	0.496	0.424	0.225
Green urban areas	0.262	0.227	0.149	0.467	0.406	0.302
Agricultural land	0.455	0.156	0.137	0.440	0.249	0.287
Natural and seminatural	0.827	0.437	0.152	0.488	0.380	0.457
LULC_avg	0.400	0.225	0.126	0.473	0.365	0.318
					ESV	0.318
Bruino						
Urban areas	0.077	0.146	0.105	0.470	0.391	0.238
Green urban areas	0.272	0.266	0.175	0.481	0.341	0.307
Agricultural land	0.378	0.168	0.190	0.463	0.395	0.319
Natural and seminatural	0.754	0.089	0.299	0.444	0.458	0.409
LULC_avg	0.370	0.167	0.192	0.465	0.396	0.318
					ESV	0.318
None						
Urban areas	0.056	0.133	0.093	0.499	0.489	0.254
Green urban areas	0.243	0.263	0.149	0.500	0.487	0.328
Agricultural land	0.402	0.215	0.214	0.496	0.484	0.362
Natural and seminatural	0.832	0.287	0.103	0.500	0.493	0.443
LULC_avg	0.383	0.225	0.140	0.499	0.488	0.347
					ESV	0.347
Settimo Torinese						
Urban areas	0.046	0.098	0.077	0.495	0.463	0.236
Green urban areas	0.237	0.248	0.141	0.495	0.456	0.315
Agricultural land	0.373	0.140	0.120	0.347	0.477	0.291
Natural and seminatural	0.723	0.336	0.150	0.486	0.488	0.437
LULC_avg	0.345	0.206	0.122	0.456	0.471	0.320
					ESV	0.320

Source LIFE SAM4CP—Actions B5 and B6

Such considerations are relevant while considering urban transformation, regeneration and rural area management policies for local planning. The character of each municipality reveals a specific ecosystemic pattern that has to be considered during the decision-making process. To highlight the latter consideration, the following schemes are designed to represent how, in the case studies in question, each ES contributes to the overall ESV quality per category of Land Use.

Further investigation was undertaken when the research group decided to look into how each single Land Use could explain how trade-offs or synergies between different services behave. To do this, a multilayered graph (see Fig. 14.2) was used applying a GIS statistical analysis of ESV distribution in the study area. The representation was used to evaluate how (i) multiple ESs are delivered for a specific category of Land Use, (ii) the configuration of ES delivery is homogeneous between particular ESs rather than heterogeneous and (iii) how each category of Land Use produces different ESV patterns.

Glancing briefly at the shape of the spider-charts, below are some considerations:

- obviously natural and seminatural areas display the better multisystemic performance of however varying character between one municipality and another;
- green urban areas perform similar values than agricultural land;
- urban areas are the only Land Use class that play a homogenous ESV distribution among the municipalities;
- while there is a synergy between higher Nutrient Retention and Habitat Quality values in natural/seminatural areas and agricultural land, such synergy is not found within the green urban areas, where low Habitat Quality values are associated with high Nutrient Retention values;

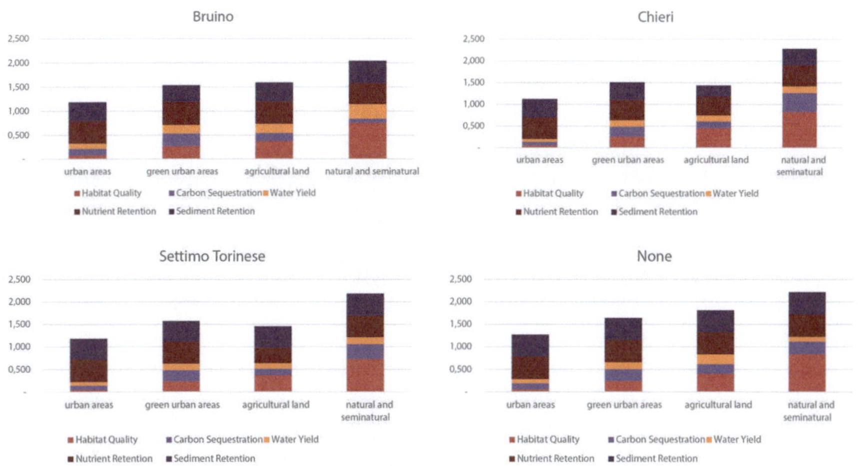

Fig. 14.2 Ecosystem Service Value composition. *Source* LIFE SAM4CP—Actions B5 and B6. Statistical elaboration of ES data with MS Excel

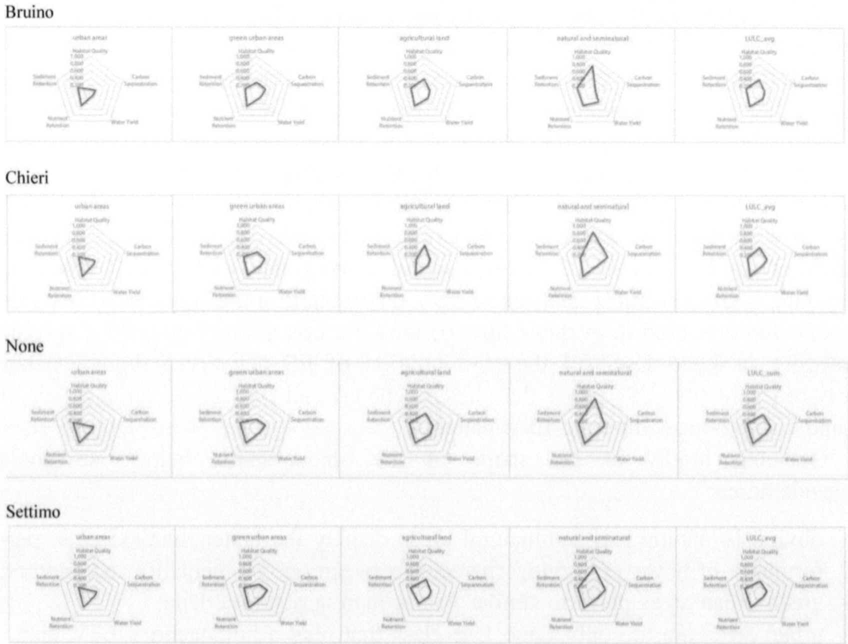

Fig. 14.3 Trade-off and synergies between Ecosystem Services across the four land categories studied. *Source* LIFE SAM 4CP—Actions B5 and B6

- the water yield value is generally low as this service is distributed in few territorial patches that are not statistically representative when compared to the average ESV value of the territory as a whole.

The charts (Fig. 14.3) are highly useful in demonstrating how every single configuration in the case study shows its level of performance on the base of a specific interaction between the context of each ES and the threats it faces. Assuming this relative perspective, the result demonstrates how policy targets for land use planning purposes must be defined by ES mapping and not merely modelling a statistical value associated with a specific category of land use. Interaction with all the ES components (topsoil and subsoil properties, land uses or land covers, vegetated biomass, morphology, climatic conditions and other management practices) generates a territorially-varied ES value that, in the case study, is a part of the same metropolitan city. This confirms the considerable variability that ES research adds to the debate around sustainability policies: not all categories of land use deliver equal benefits and, moreover, the distribution of benefits is unequal over different land use configurations; therefore, no aprioristic assumption should be made while evaluating land use change assessment.

14.3 Conclusions

As we were expected, none of the case studies of LIFE SAM4CP show effective use of knowledge in the sense that ecosystem service knowledge would have been the "missing piece" of information needed to make decisions (Saarikoski et al. 2018). This is quite understandable because knowledge as such cannot serve as an impartial arbiter in such decision-making situations.

But the public and private stakeholders involved felt that the mapping and valuation research process resulted in the changed understanding of the importance of the different classes of land use, of green spaces in urban areas and green infrastructures.

Since one of the challenges of LIFE SAM4CP concerns the limiting of land consumption and in particular that agricultural, one of the themes addressed by the Variants of the Local Plan concerns the reversibility of land uses. A very topical issue also due to the economic and real estate crisis and a tax on building areas that penalizes the "waiting land rent".

In this perspective, the cases of LIFE SAM4CP relate to experiences of municipal land use planning that aim to reduce planned expansion in the agricultural territory by Land use plan variants aimed at removing the settlement forecasts, downgrading them to agricultural uses. The political motivations that accompany these procedures refer to objectives of regeneration of the urbanized territory, of reduction of land consumption and contrast to climate change. In the cases of Chieri and None, the urban spatial planning procedure was preceded by governance actions through stakeholder consultation practices based on the collection of interest—by the public announcement—to the urban reclassification of potentially urbanized areas that have not yet been implemented (and which therefore constitute the so-called "planning residue"). It is impressive to note that in the case of Chieri, the *Structural Land Use Variant* "cuts" 127 ha of forecasts present in the current *Land Use Plan*, and, compared to the current configuration of the territory, adds 30 ha of areas to serve the new agricultural/fruitive park.

The data collected during the project is insufficient to exclude the possibility that similar patterns should be founded in municipalities sharing similar characteristics with those of the project evaluation but, at the same time, the assessment is detailed enough to confirm that land use regulation requires the input of highly diversified data. Notably, the decision-making process requires the strong support of environmental data to obtain sustainability; nevertheless, the specificity of the data often does not fulfill the need for simple, selected and easy-to-understand information.

The latter point builds a connection to what the SEA has become within the legal framework of the 2001/42/CE Directive. In the vast majority of cases, the construction of the plan is not aided by SEA indicators and considerations, as the scope of such an assessment is to assist the planning process providing constant feedback between land use plan alternatives and their real-time evaluation through different environmental indicators.

In the SAM4CP research project, the research group learnt how to address real SEA implementation contributions using ESV as a proxy for land use sustainability. The experimentation in the Co-planning Conferences of Bruino, Chieri, None (February–December 2017) and Settimo Torinese (March 2018) showed that the various administrative subjects who participated (of the Region and Metropolitan City), in particular the environmental authorities, welcomed and interest in this methodology (mapping and biophysical evaluation of soils in different time scenarios) to assess the sustainability of the hypothesis of land use modification.

The combination of soil quality state indicators and future environmental pressures (e.g., alternative land use in suburban areas rather than substituted land use within the anthropic environment) generated by potential transformation contributed significantly to understand the sustainability of the land use modification. It also helped the drafting of policy responses that take into consideration limitation, mitigation or compensation measures (European Commission 2012) in restoring ES conditions.

Within this operational perspective, the urban plan can experiment with the innovation of the relationships between design activity and knowledge system that should guide and legitimize the choices of territorial governance within a dimension that can simultaneously be analytical-evaluative and soil design-based.

In this sense, the planning method renews its ability to substantiate the pursuit of the general interest in supporting choices oriented to the sustainability of environmental resources.

References

Alam, M., Dupras, J., & Messier, C. (2016). A framework towards a composite indicator for urban ecosystem services. *Ecological Indicators, 60,* 38–44.

Antrop, M. (2004). Landscape change and the urbanization process in Europe. *Landscape and Urban Planning, 67,* 9–26. https://doi.org/10.1016/S0169-2046(03)00026-4.

Barbieri, C. A. (2017). Legiferare sul contrasto al consumo di suolo, ma come e chi? In A. Arcidiacono, D. Di Simine, F. Oliva, S. Ronchi, & S. Salata (Eds.), *La dimensione europea del consumo di suolo e le politiche. Rapporto CRCS 2017* (pp. 247–250). Rome: INU Edizioni.

Barbieri, C. A., & Giaimo, C. (2015). A new model of institutional governance for new planning policies in Italy. *Urbanistica, 153,* 92–97.

Butsic, V., Shapero, M., Moanga, D., & Larson, S. (2017). Using InVEST to assess ecosystem services on conserved properties in Sonoma County, CA. *California Agriculture, 71*(2), 81–89. https://doi.org/10.3733/ca.2017a0008.

Clerici, N., Paracchini, M. L., & Maes, J. (2014). Land-cover change dynamics and insights into ecosystem services in European stream riparian zones. *Ecohydrology and Hydrobiology, 14,* 107–120.

European Commission. (2012). *Guidelines on best practice to limit, mitigate or compensate soil sealing*. Commission Staff Working Document, SWD (2012), 101 final, Brussels. http://ec.europa.eu/environment/soil/pdf/soil_sealing_guidelines_en.pdf. Accessed February 22, 2018.

Grêt-Regamey, A., Altwegg, J., Sirén, E. A., van Strien, M. J., & Weibel, B. (2017). Integrating ecosystem services into spatial planning—A spatial decision support tool. *Landscape and Urban Planning, 165,* 206–219. https://doi.org/10.1016/j.landurbplan.2016.05.003.

Kaczorowska, A., Kain, J.-H., Kronenberg, J., & Haase, D. (2016). Ecosystem services in urban land use planning: Integration challenges in complex urban settings—Case of Stockholm. *Ecosystem Services, 22,* 204–212. https://doi.org/10.1016/j.ecoser.2015.04.006.

Langemeyer, J., Gómez-Baggethun, E., Haase, D., Scheuer, S., & Elmqvist, T. (2016). Bridging the gap between ecosystem service assessments and land-use planning through Multi-Criteria Decision Analysis (MCDA). *Environmental Science and Policy, 62,* 45–56. https://doi.org/10.1016/j.envsci.2016.02.013.

Laurans, Y., Rankovic, A., Billé, R., Pirard, R., & Mermet, L. (2013). Use of ecosystem services economic valuation for decision making: Questioning a literature blindspot. *Journal of Environmental Management, 119,* 208–219. https://doi.org/10.1016/j.jenvman.2013.01.008.

Maes, J., Egoh, B., Willemen, L., Liquete, C., Vihervaara, P., Schägner, J. P., et al. (2012). Mapping ecosystem services for policy support and decision making in the European Union. *Ecosystem Services, 1,* 31–39. https://doi.org/10.1016/j.ecoser.2012.06.004.

Mubareka, S., Koomen, E., Estreguil, C., & Lavalle, C. (2011). Development of a composite index of urban compactness for land use modelling applications. *Landscape and Urban Planning, 103,* 303–317. https://doi.org/10.1016/j.landurbplan.2011.08.012.

Partidario, M. R., & Gomes, R. C. (2013). Ecosystem services inclusive strategic environmental assessment. *Environmental Impact Assessment Review, 40,* 36–46. https://doi.org/10.1016/j.eiar.2013.01.001.

Pulighe, G., Fava, F., & Lupia, F. (2016). Insights and opportunities from mapping ecosystem services of urban green spaces and potentials in planning. *Ecosystem Services, 22,* 1–10. https://doi.org/10.1016/j.ecoser.2016.09.004.

Salata, S. (2017). Land use change analysis in the urban region of Milan. *Management of Environmental Quality: An International Journal, 28*(6), 879–901. https://doi.org/10.1108/MEQ-07-2016-0049.

Salata, S., & Gardi, C. (2014). From quantitative to qualitative analysis of Land-Take. The application of a Composite Indicator for targeted policies of Land Take reduction. *City Safety Energy, 1*, 15–31. http://dx.doi.org/10.12896/cse20150010041.

Salata, S., Garnero, G., Barbieri, C., & Giaimo, C. (2017). The Integration of ecosystem services in planning: An evaluation of the nutrient retention model using InVEST software. *Land, 6,* 48. https://doi.org/10.3390/land6030048.

Saarikoski, H., Primmer, E., Saarela, S.-R., Antunes, P., Aszalós, R., Baró, F., et al. (2018). Institutional challenges in putting ecosystem service knowledge in practice. *Ecosystem Services, 29,* 579–598. https://doi.org/10.1016/j.ecoser.2017.07.019.

Schröter, M., Remme, R. P., Sumarga, E., Barton, D., & Hein, L. (2015). Lessons learned for spatial modelling of ecosystem services in support of ecosystem accounting. *Ecosystem Services, 13,* 64–69.

Setälä, H., Bardgett, R. D., Birkhofer, K., Brady, M., Byrne, L., de Ruiter, P. C., et al. (2014). Urban and agricultural soils: Conflicts and trade-offs in the optimization of ecosystem services. *Urban Ecosystems, 17,* 239–253. https://doi.org/10.1007/s11252-013-0311-6.

Silvennoinen, S., Taka, M., Yli-Pelkonen, V., Koivusalo, H., Ollikainen, M., & Setälä, H. (2017). Monetary value of urban green space as an ecosystem service provider: A case study of urban runoff management in Finland. *Ecosystem Services, 28,* 17–27. https://doi.org/10.1016/j.ecoser.2017.09.013.

Tallis, H., Ricketts, T., & Guerry, A. (Eds.). (2011). *InVEST 2. Beta user' s guide*. Stanford: The Natural Capital Project.

Turkelboom, F., Thoonen, M., Jacobs, S., & Berry, P. (forthcoming). In M. Potschin & K. Jax (Eds.), *OpenNESS Reference Book*. EC, FP7 Grant Agreement no. 308428. Available via: www.openness.project.eu/library/reference.

Carolina Giaimo Ph.D. in Spatial Planning (IUAV-Venice), is an Assistant Professor in Urban Planning at the DIST Department of the Politecnico di Torino, where she researches innovative perspectives in the field of urban regeneration and sustainable urban and territorial government. She is involved in the European project LIFE SAM4CP since 2015. Member of the Ecosystem Service Partnership (ESP) since 2015, member of the Italian National Institute of Planning (INU) since 1990 and member of the National Board since 2016, she has been Vice President of the INU Chamber for the Regions of Piedmont and Aosta Valley since 2004. Member of the Editorial Board of the Journals Urbanistica and Urbanistica Informazioni and author of many national and international publications.

Carlo Alberto Barbieri is a Full Professor in Urban Planning at the DIST Department of the Politecnico di Torino. He has more than 30 years of teaching experience in Bachelor Degree, Master of Science Degree and Doctorate Courses. He is the Scientific coordinator for DIST of the European project LIFE SAM4CP. He has been national Vice President of the Italian National Institute of Planning (INU) and since 2013 has been Vice President of the INU Chamber for the Regions of Piedmont and Aosta Valley. Member of the National Superior Council of Public Operas of Italian Republic, he is designer plans and developments of several towns in Italy (Piedmont, Liguria and Campania Regions). Author of many national and international publications and studies on urban planning and territorial government.

Stefano Salata Ph.D. in Territorial Government and Urban Design at the DAStU Department of the Politecnico di Milano, is involved in research activities on land use change and ecosystem services assessment. Since 2015 is involved in the project LIFE SAM4CP at the DIST Department of the Politecnico di Torino, where he works as technologist at the R3C-Responsible Risk Resilient Centre since 2018. He is a member of Ecosystem Service Partnership (ESP) and of the National Research Centre on Land Take—ITALY (CRCS). Co-editor of Italian National Reports on land take (2010, 2012, 2014, 2016 and 2017) of CRCS, member of the Editorial Board of the Journal of Architecture, City and Environment and author of many international publications.

Chapter 15
Spatial Tools to Renew the Planning Paradigm Through Metropolitan Agriculture. The Case Study of Turin (Italy)

Enrico Gottero and Claudia Cassatella

15.1 Introduction

Currently UA is no longer a residual activity. It is a very heterogeneous phenomenon, where the UA intensity and form depends on the country under consideration: geographical location, socio-economic context, the system of governance, public policies, cultural practices, environmental features, morphology, etc. These are conditions that increase or reduce diversity and build a complex set of international practices (Hsin 1996; Bhatt et al. 2005; Vidal 2009). According to the most recent research experiences (see, for example, Lohrberg et al. 2016), UA includes urban spaces that farmers continue to cultivate in an attempt to attract the interest of the city market, where the citizens grow their own vegetable garden, but also greenfield sites, as well as brownfields and unbuilt areas (parks, forests, cultivated areas) (Branduini 2008).

At the same time, several dimensions of UA include a complex system of values and functions. UA can respond to a variety of social, economic, cultural and environmental objectives. It is the breeding ground for increasing and strengthening social aggregation (ethnic and generational), for improving the quality of life (well-being, physical and mental health), for educational reasons, subsistence for families, protection of territory and soil, etc. Similarly, UA is an important environmental asset contributing to biological cycles (air, water, soil). In other terms UA provides an opportunity to use and manage residual and public green spaces (as

E. Gottero (✉)
IRES Piemonte, 10100, Torino, Italy
e-mail: e.gottero@ires.piemonte.it

C. Cassatella
DIST, Interuniversity Department of Regional and Urban Studies
and Planning, Politecnico and Università di Torino, 10125 Turin, Italy

E. Gottero (ed.), *Agrourbanism*, GeoJournal Library 124,
https://doi.org/10.1007/978-3-319-95576-6_15

in the case of urban agricultural parks), but also a testing tool for the quality of agricultural production and for the efficiency in linking material flows and shortening the supply chains. UA positive externalities attributed through the benefits received by citizens, reside also in the new demand for common goods and services, such as quality and typicality of foods, as well as in the growing phenomenon of rural tourism (Donadieu 2004; Tempesta 2011). UA can also contribute to improve recreational and leisure facilities and—through agricultural techniques and procedures (tilling, grazing, harvesting, recycling, etc.)—it can help to restore and maintain the identity of traditional places, as well as the aesthetic value of agricultural landscapes (Caspersen and Olafsson 2010; Laviscio 2014). UA has a high knowledge and innovation potential. Therefore it can contribute to give strong impulse to the whole agricultural sector of most European cities because it also offers benefits for *«social inclusion and job creation on a low threshold»* (Lohrberg 2013, p. 1).

Although acknowledged both in the scientific and institutional context, this complex system of values at stake is often unknown or overlooked. Currently the policies framework of several European cities, do not seem to meet these new social issues. The lack of structured cognitive frameworks of UA forms and their interaction with complex governance systems, in order to support decision-making in fostering UA and to carry out strategies traced by previous plans or programs, is a common issue for many cities. In addition, some European support instruments, such as the *Rural Development Policies*, have proved to be often ineffective and not very sensitive in terms of actions related to agricultural landscapes in the urban and periurban areas (Gottero 2015). Although the *Common Agricultural Policy* (CAP) also aims to support farmers and agriculture in urban areas, in most cases (as in Italy), it has directed most of its principles and goals specifically at farmers in "the countryside" and "in rural areas" (Lohrberg et al. 2016).

Thus, these complex issues underline as an approach based on rethinking UA has become a necessity. Adopting *site-specific* solutions, this new approach will can change the focus from urban development to farmland planning (Branduini 2008), where regions, cities and civic society establish new types of public policies and new forms of governance dealing with UA.

In this regard the recent research project "Urban agriculture innovating Torino Metropolitan Area (TMA). Tools for governance and planning of a complex system"[1]—supported by ISI Foundation, CRT Foundation and IRES Piemonte in the context of Lagrange scholarship (2015–2016)—has been conducted. The TMA represents an evolving reality, where, in the last decade, several projects on urban gardening and farming have been launched. In this chapter the authors show some results of this research. In the first section the methodological choices for the construction of a cognitive and interpretative framework for UA in the TMA, will be illustrated, while in subsequent sections the main findings with some elements of

[1]It has been carried out with scientific supervision of Claudia Cassatella (DIST) and Stefano Aimone (IRES Piemonte).

discussion will be tackled. The last section will be dedicated to limits, opportunities and open questions for research topics and for possible research perspectives.

15.2 Methodological Approach

15.2.1 Description and Evolution of the Study Area

The case study area is located in Middle-Western Piedmont (Italy) and is strongly polarised by the Turin urban conurbation. The territories included in this context coincide with 38 municipalities, namely an area in which today there are about one and a half million inhabitants, in a total surface area of 858.67 km^2. Among the most significant structuring factors are the Turin plain and the presence of numerous watercourses such as the Po, Sangone, Dora and Stura. A hilly area (mainly woodland) extends to the east, while to the north of Turin, the extensive farming area is mainly oriented towards grassland and rotation crops. To the south, the Stupinigi plain is important, as is the cereal-growing and the fruit and vegetable crops of Poirino, while to the west, the presence of the morainic amphitheatre of Rivoli-Avigliana represents a marginal agricultural landscape strongly influenced by the proximity of the city of Turin. Among naturalistic emergencies, the presence of protected areas and *European Sites of Community Importance* (SCI) (La Mandria, Stupinigi and the Turin hills), as well as the Po river system (*UNESCO Biosphere Reserve*) are very significant. These areas are also particularly rich in assets including farmsteads and works of various kinds that are typical of the traditional rural landscape. Among the structuring factors we would mention the Savoy residences including Venaria, Rivoli, Stupinigi, and settlement systems developing along the Turin-Lanzo and Turin-Val di Susa highways (Piedmont Regional Authority 2015).

However, these are territories that have been the subject of very contrasting dynamics in recent decades (Table 15.1). As for the population within the TMA, the demographic balance in the period between 1991 and 2016 shows a substantial fall of −0.64% per annum—based on ISTAT database.[2] The result that emerges is also significant, if from the series we exclude the number of inhabitants of the city of Turin. In this case, the balance is actually positive, above 9% over the same period. It is also interesting to observe how growth is mainly concentrated in the green belt of Turin, while the decrease is primarily a phenomenon attributable to the most crowded municipalities of the first suburbs. The change in land use and land cover between 2000 and 2012—based on *Corine Land Cover* (CLC)—shows that the areas most affected by the changes concern especially the part to the east of

[2]In this regard see: Osservatorio Demografico Territoriale del Piemonte (DemOs)–http://demos.piemonte.it/ (accessed: 13/12/2016) and i.STAT—banca dati delle statistiche—http://dati.istat.it (Accessed: 13/12/2016).

Table 15.1 The dynamics of the study area

Description	Amount	Source
Inhabitants (1991) (n.)	1,584,766	ISTAT
Inhabitants (2016) (n.)	1,574,673	
Urban area (2000) (ha)	22,569.03	CLC
Urban area (2012) (ha)	24,595.04	
Green urban areas (2000) (ha)	925.15	
Green urban areas (2012) (ha)	1,234.76	
Agricultural areas (2000) (ha)	50,037.86	
Agricultural areas (2012) (ha)	48,591.59	
Forestal areas (2000) (ha)	10,437.89	
Forestal areas (2012) (ha)	9,233.46	
Farms (2000) (n.)	3,349	ISTAT
Farms (2010) (n.)	2,422	
Municipal UAA (2000) (ha)	31,454.98	
Municipal UAA (2010) (ha)	30,368.05	

Source Reworking of the authors from: Gottero (2016)

Turin and, in particular, the agricultural and wooded areas to the north, which are also affected by phenomena such as abandonment and depopulation. Analysis of the rate of land use variation shows that in this period the increase in man-made areas corresponds to a significant decrease in forestal and cultivated areas, as well as an exponential growth of evolving wooded and shrubby areas, i.e. that can result from degradation or recolonisation of non-forestal areas. The urban sprawl that has affected the entire metropolitan area of Turin, especially in the period under review, is particularly evident even from the analysis of the changes in urbanised areas. This has had a significant impact on the rural areas of the TMA: the transformation of these for productive and residential use, in addition to the development of infrastructure, has gradually triggered a loss of traditional identity elements, as well as a consequent crop variation. The abandonment of rural areas and the most recent changes in land use that have characterised the TMA can also be traced back to other phenomena closely related to urban farming activities. In this regard, if we observe the surveys conducted through the ISTAT agricultural censuses (2000–2010),[3] the data are apparently conflicting. Although the number of farms has declined substantially over this period, i.e. close to 30% in value, by contrast the *Utilised Agricultural Area* (UAA)—namely including that outside the TMA—is increasing. This question probably corresponds to the increase in the average size of farms (from 10 to 15 ha), which is offset by a substantial decrease in the municipal UAA (approximately 13%), i.e. the area under crops within the TMA.

[3]In this regard see: ISTAT—Censimento dell'agricoltura 2010—http://dati-censimentoagricoltura.istat.it/ (Accessed: 13/12/2016).

15.2.2 The Construction of a Cognitive Framework of UA in the TMA

The literature has repeatedly and in various contexts reaffirmed how, to date, urban farming can be considered a highly complex phenomenon, which embraces multiple purposes, constituting a heterogeneous activity that can take on different forms and dimensions. By adopting this approach, in the following pages we will analytically reconstruct the overall cognitive framework of the TMA, in order to bridge the current shortcomings that omit not only the places where farming is practised in the city, but also the different forms of administration and/or management of urban agricultural spaces.

To this end, the research has adopted the "by purpose" classification proposed by Lohrberg et al. (2016), which distinguishes two types of urban agriculture: *professional* (urban farming) and *not professional* (urban gardening). The former is supported by economic-productive purposes, as well as being oriented to the sale of products and services. Adopting the approach described by Simon-Rojo et al. (2016), this form of UA corresponds to a dual distribution, i.e. *Oriented Urban Farming* (OUF) and *Not Oriented Urban Farming* (NOUF). The OUFs, i.e. educational, social, recreational and other farms, use business models that intentionally seek to exploit proximity to the urban environment by offering local and regional agricultural services and/or products. These are farms which apply different diversification strategies in order to meet urban demand, especially in the recreational, tourist and environmental fields. Multifunctionality in urban areas therefore translates into different types of urban farming, single or combined, oriented not only towards the provision of local services (recreational, therapeutic, social, etc.), but also linked to environmental benefits and food quality (Zasada 2011). Despite their proximity to the urban environment, NOUFs maintain conventional agricultural activities, where production is generally oriented towards satisfying the national or international market.

In the present case, the research first of all attempted to carry out a functional breakdown by working on the uniform sample contained in the *regional data-warehouse* of the *Anagrafe Agricola Unica del Piemonte*[4] (AAU) (Piedmont Farm Register). This register contains data on agricultural and agri-food holdings, as well as those of different actors initiating procedures in agriculture and rural development. The data contained in the register were first of all cleared of the values not present in ISTAT's agricultural census of 2010, an operation that standardised the data set and adjusted the analysis sample to a value of 2,173 units (of which 2,151 geo-localisable), a number that was closest to the number of farms registered in the census (2,422). Using these data as a basis, the OUF's were then geo-located using

[4]In this regard see: Anagrafe Agricola Unica (AAU)—datawarehouse—http://www.sistemapiemonte.it/cms/privati/agricoltura/servizi/367-anagrafe-agricola-unica-data-warehouse (Accessed: 13/12/2016).

Fig. 15.1 The Oriented Urban Farming in the TMA. *Source* Reworking of the authors from: Gottero (2016)

the *QGIS geocoding tool*.[5] The classification of professional urban farming "by purpose" was subsequently directed at distinguishing between urban market-oriented farming activities (OUF) and non-urban market oriented farming activities (NOUF). Of these 118 are OUF (Fig. 15.1), while the remaining 828 are

---

[5]In this regard, it should be pointed out that the AAU figure refers to the registered office of the farm, which in most cases corresponds to the location it operates from. However, in order to avoid problems of incorrect geo-location, hypothesising as plausible only farm locations outside densely built-up areas, the data were purged through overlay operations between detailed information of each individual farm and the spatial data of the land coverage deriving from the CLC (IV level, 2012), namely excluding all artificially modelled territories except artificial non-agricultural green areas. The exclusion of these data reduced the survey sample from 2,157 geo-localisable farms in the AAU to 946 units, i.e. those outside the urbanised area. At the same time, this operation results in a drastic reduction of the sample, especially with regard to OUFs, which go from a real value of 211 to 118 farms. Although not optimal, the decision to exclude such farms that have their business centre, i.e. the place where they offer goods and services, in another municipality, makes it possible to operate on a sample of farms that may not be exhaustive but is still sufficiently reliable.

NOUF. The former correspond to holdings that perform at least one of the complementary activities—those that ISTAT calls related remunerative activities such as agritourism, educational farming, biologic production, recreational and social activities, etc—while the others have adopted a conventional scheme that does not support any of these complementary activities and tends to use an undiversified business model. For this analysis, we have also resorted to the use of data from the most recent surveys conducted on agritourism activities, the farms present in the regional list of educational farms,[6] as well as the data extracted from the biological process (alphanumeric data provided by CSI Piemonte, updated to 11/11/2015).

In contrast to the above, family or amateur urban agriculture (*Urban Gardening*) is often practiced by people or families as a non-commercial activity and for self-support, or with predominantly social aims. Within the TMA, as in other metropolitan situations, the gardening phenomenon is variegated and fragmented, even if it is now widely developed over the whole territory of plain and hills. However, the knowledge gaps particularly evident in the case of the TMA have so far prevented understanding of a multifaceted phenomenon that often takes on "polychrome" connotations that are difficult to classify. Moreover, there is no exhaustive database of vegetable gardens at present, but rather individual experiences and attempts at mapping and classification. For these reasons, in order to build a complete cognitive picture, the analysis conducted on the TMA has operated through the photo-interpretation method, that is to say the interpretation of images —in this case with the help of AGEA's aerial photographs (2015)—and the identification through GIS tools, of pinpoint or area elements (Fig. 15.2).

15.2.3 The Construction of an Interpretative Framework of UA in the TMA

The need for evaluation tools and frameworks to address ecosystem service issues such as, for example, food safety and nutrition, cultural and environmental capital, citizens' well-being and quality of life, in an effective and inclusive way, is now considered to be an urgent issue, especially in metropolitan regions (UN Habitat 2015). In the following pages we will therefore look into the various opportunities and the current basic ecosystem functions attributable to urban and peri-urban agriculture of the TMA, with the help of indicators. These are tools developed on the metropolitan scale in order to build a set that can be used for multiple purposes such as context analyses, *Strategic Environmental Assessments* (SEAs) as well as assessments of potential and specific requests for multi-fund financing.

In order to describe, analyse and measure the UA in the TMA, research has adopted an ecosystem approach that enables us to verify and establish the

[6]In this regard see: http://www.regione.piemonte.it/agri/politiche_agricole/edu_aliment/fattorie.htm (accessed: 26/07/2016).

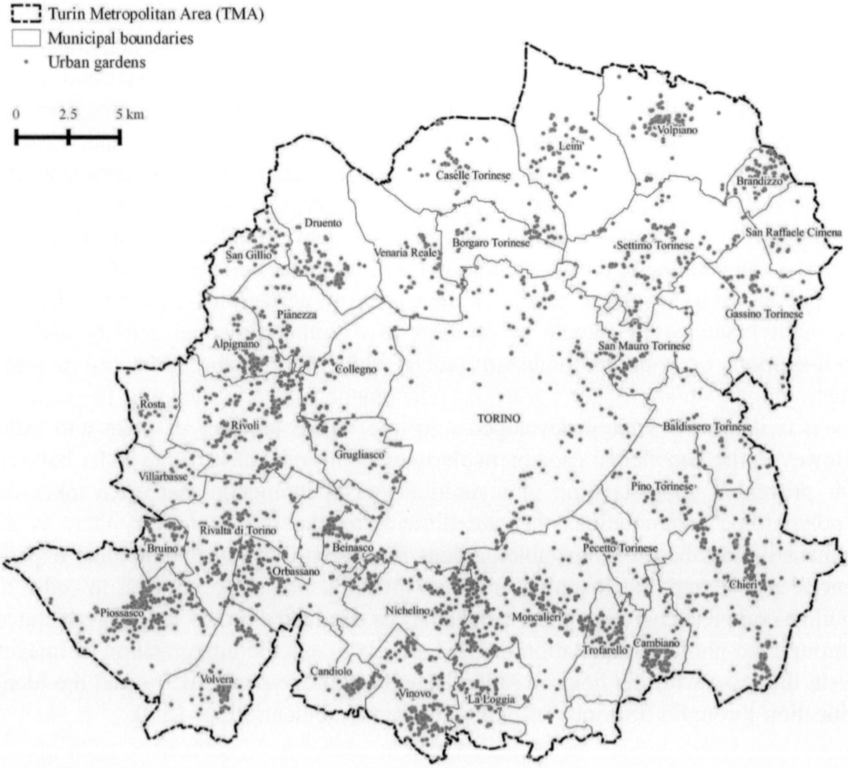

Fig. 15.2 The urban gardens in the TMA. *Source* Reworking of the authors from: Gottero (2016)

contribution of agriculture in more than one direction and orientation. This approach thus makes it possible to give shape to the multiple dimensions of the metropolitan agricultural landscape, give tangible form to seemingly immaterial ones and understand the complexity of a real socio-ecological system (Willemen et al. 2008; Termorshuizen et al. 2009). The proposed scheme for the TMA, based on the interpretative model proposed by Ingersoll et al. (2007), first attempts to determine a system of fundamental relationships based on essential functions for the metropolitan area belonging to the main ecosystem classes (provision, regulation, habitat, cultural), and on the potential opportunities that can derive from them (nutritional, ecological, social, educational, recreational, therapeutic, cultural, control). The set of key indicators/indices (in this regards see also Gottero 2016), selected from the existing literature and developed for the TMA, aims to highlight the phenomena that contribute to the development of a plurality of ecosystem classes and services (Table 15.2).

Table 15.2 The set of key indicators and indices for the TMA

Indicator/index	Functions							
	Alimentary	Maintenance	Ecological	Social	Educational	Recreative	Therapeutic	Cultural
Certified products	•							•
Food production	•							
Food provision	•							
Supposed food demand	•							
Percentage of the municipal area dedicated to the public green area			•					
Soil sealing index	•		•					
Availability of public green space per capita				•	•		•	
Cost of urban green areas management				•				
UAA per capita	•	•						
Usability of urban oriented farm				•		•		
Accessibility to urban oriented farm				•		•		
Recreational areas per capita						•	•	
Population with access to green recreational areas						•		
Urban agri-tourism					•	•		
Damaged landscapes			•					
Vulnerability index of metropolitan landscape		•	•					•
Socio-economic discomfort				•				
Supposed inclusion demand				•				

Source Gottero (2016)

15.3 Results

15.3.1 Cognitive Framework

As far as professional urban agriculture, as a prevalent form in the TMA, is concerned it consists in the main of farmers and horticulturists, individuals or associates in cooperatives, family-run farms and agri-food companies. OUFs are therefore in a clear minority, i.e. around 12% of the total, and with an average standard production lower than non-oriented production. They cover about 20% of *Total Agricultural Land* (TAL) and involve 15% of UAA, while they are on average larger than NOUFs. These are mainly agri-tourisms (24), educational farms (17) and social farms (13), while there are only 5 organic farms. In addition, the OUFs received around 10% of the CAP contribution in the period 2007-2013. The OUFs are mainly located in the south-east of the TMA and on the western side of Turin. These are in the main holdings specialising in horticulture (28%), arable crops (18%) and herbivores (15%), while less present (<5%) are those dedicated to permanent crops and mixed livestock (Table 15.3). OUFs are on average larger than NOUFs and generally include specialised arable, fruit and vegetable and cattle farms. The analysis also showed that non-conventional farms received less significant public support than others. However, there are still many farms that have not adapted the economic-productive model to the urban market, especially in the south-west and north-east quadrant, although they coincide with municipalities with a positive demographic balance, areas of particular agronomic interest and with a high density of historical evidence of rural territory, as well as with sites of particular naturalistic value.

On the not professional side, the research showed the presence of countless, diversified experiences of urban gardening which, contrary to urban farming, are mainly located on the south side and, in particular, in the south-west quadrant, almost as if to offset the lack of the component attributable to urban farming. However, it is a small number when compared to the number of inhabitants and the growing demand for new cultivatable urban spaces, with different purposes (social, food, recreational, etc.), as well as if correlated to the potential contribution of vegetable gardens in terms of greening the urban environment. The TMA survey

Table 15.3 Urban farming in the TMA

	OUF	NOUF
Number	118	828
UUA (ha)	2,203.83	13,035.94
TAL (ha)	3,437.75	13,946.74
Average UUA per farm (ha)	19.68	17.15
CAP contribution (€) (2015)	5,625,992.75	49,039,459.34
CAP contribution per farm (€) (2015)	47,677.90	59,226.40

Source Reworking of the authors from: Gottero (2016)

Table 15.4 Urban gardens in the TMA

UG features	Amount
Number of urban gardens (UGs)	3,809
Average size (mq)	1,160.17
Number of UGs demand	1,269
UGs surface per inhabitant	6.52
Number of UGs per inhabitant	0.0068

Source Reworking of the authors from: Gottero (2016)

showed the presence of 3,809 urban gardens of which about 70% are kitchen gardens, namely private vegetable gardens located near single homes, and 20% are family gardens in non-built-up or agricultural areas. Less significant in numerical terms are the public social gardens (i.e. gardens assigned and formalised), community gardens, as well as the spontaneous and sometimes illegal gardens along watercourses and in the city-fringe. The total area occupied by urban gardens is about 485 ha, with an average size close to 1,200 m^2, i.e. about 6.5 m^2 per inhabitant and less than one garden per inhabitant. Family gardens and spontaneous gardens are larger, 2.2 million and 1.1 million square metres respectively, while social gardens and kitchen gardens seem to be less significant (Table 15.4). Specifically, when expressed in terms of area, the latter cover only 14% of the total area, whereas in terms of units they are a majority. As for the localisation of urban gardens and their distribution in the metropolitan area, GIS analysis also showed a strong concentration of urban gardens especially on the southern slope, in the south-west quadrant, and in the northeast end of the TMA. The areas, on the other hand, that have large zones without urban gardens are located on the north-west border, in the intraurban areas of Turin, as well as in the hills between Gassino and Baldissero.

15.3.2 Interpretative Framework

The key to interpreting the global significance of urban agriculture lies precisely in its current functions and ecosystem potential. The benefits provided to mankind are not limited to the mere food supply component, but are wider in scope, although the food issue in the TMA is certainly one of the most significant aspects. In this regard, the analysis carried out has highlighted a very complex agro-food system: in terms of quality, the "certified" food in the TMA is produced mainly in low yield per hectare agricultural areas—think of the low production municipalities in the south-west quadrant—except for the hills south-east of Turin. In terms of energy, the southern side of the TMA makes the highest calorie contribution, although of low quality and with demand particularly restrained. The presumed demand for food is thus largely satisfied in the TMA, although there are still some municipalities where the supply/demand balance remains negative (especially in Turin).

As for the contribution in terms of greening and improving the quality and salubrity of the places, metropolitan agriculture in the TMA can contribute to increasing the share of open and undeveloped green spaces per inhabitant and, as a result, to improving the quality of life of citizens, especially in the municipalities of the southern quadrants with a high UAA per capita and low availability of public green areas. Furthermore, research has highlighted the high cost of managing public green areas, especially in the south-west quadrant, in correspondence with areas where there is little soil sealing and with a significant number of farms.

As regards the fruition and recreation sphere, the TMA evaluation has shown above all that the accessibility and usability of oriented farms, i.e. those that are structured to meet urban needs, is deficient, especially in the south-east quadrant, while the opposite is often the case in the other southern sector. The ring around Turin and its significant number of recreational areas per capita seems also to correspond to the hemicycle that runs from north-west to south-west, i.e. an area with low practicability and use of OUFs also due to a limited cycle-pedestrian infrastructure. These are also recreational areas where, especially in the section from east to south-west, the number of inhabitants who can access them is small compared, for example, to the municipalities in the north-west quadrant. Agritourism activities, on the other hand, seem evenly distributed to the west of Turin and on the hill-plain area to the east and south-east, contrary to a very limited offering in the south-west quadrant, as well as in the north-east sector, which at the same time is difficult to access. Contrary to the above, the socio-cultural contribution of urban farming is based on verifying the fragility of the metropolitan area, understood in terms of maintaining or creating the conditions that affect the quality of life and well-being of citizens (environmental quality, identity and historical-cultural heritage, social inclusion, physical and mental health, work, income, etc.). According to the study, the citizen of the TMA is therefore living in particularly unfavourable environmental circumstances, especially in the north-eastern quadrant and on the southern side, areas where the current human footprint and the risk of losing not only agricultural land, but also material memory due to new settlements, is particularly significant. At the same time, the social and economic hardship recorded in the TMA is particularly significant in the city itself and in the first suburbs of Turin, while in the semicircle that wraps round this city core from east to west, in correspondence with the municipalities that are less "damaged", as well as being "greener" and more "recreational", the malaise seems to be diminishing.

The set of indicators we have described briefly are appropriate and sufficient for describing the main dimensions of the ecosystem that gravitates around the urban farming system. However, they lack those spatially defined indications and methods for their implementation which are indispensable for the correct, informed planning of agro-urban spaces, as well as for the implementation phase of strategies. In this direction, as a further element of synthesis and decision-making support, the research has developed an additional device that finds an effective new solution in the union of urban planning and sector planning tools. It is therefore a meta-project

WHERE	HOW	
	Objectives	Measures
	– Strengthen high quality food production; – Meet the demand for local food – Promote food equity (access and food quality)	1 F - Support for the creation of direct sales points;
		3 F - Support for the professional agricultural reconversion of brownfield areas (soilless cultivation, vertical farming);
		4 G - Support for the social agricultural reconversion (urban gardening) of brownfield areas (soilless cultivation)
		8 F - Support to farmers for adaptation to alternative food networks and new urban sales systems;
		12 F - Support for the creation of family gardens for self-support (allotment public space)

Fig. 15.3 The "Where-How" model: an example of a Map-based indicator (Municipalities with negative food balance) and the meta-project tool application. *Source* Reworking of the authors from: Gottero (2016)

that combines indications and criteria of spatial targeting—map-based tools derived from numerical evidence of indicators—with objectives, intervention priorities (farming and gardening) and a package of n. 25 measures (multifund) specific to metropolitan areas (at the farming and gardening level), i.e. the permitted interventions and eligible areas (see, as an example, Fig. 15.3). Tools on which to base decisions, identifying where to support targeted interventions, how to implement strategies already identified by urban planning and intercept resources from *European Structural and Investment Funds (SIF)*. Elements called for, inter alia, by the already mentioned UN Habitat conference.

Generally speaking, the results of applying this model to the Turin case study have highlighted the relevance and urgency of an intervention specifically oriented towards urban agriculture, especially in the municipalities that most gravitate round Turin (especially the core area of Turin), as well as on situations distant from the urban centre but strongly characterised by the presence of agriculture (Chieri, Leinì, Piossasco, Bruino). Finally, there is evidence of a ring of municipalities with average territorial priorities, which from the north continue clockwise to the west. The municipalities on the northern border seem to be of lower priority (Fig. 15.4).

15.4 Discussion

The results that have emerged from the research carried out as part of the Turin case study have highlighted some important elements, especially with regard to the validity and applicability of the method and indicators tested. Generally speaking, the results have proved easy to replicate, in different institutional contexts, also as a basis for supporting the sector policy decision-making system. It is also easy to transfer this model of analysis to other metropolitan contexts, considering how widespread and accessible the sources and databases used for the construction are. As for the cognitive framework, although not exhaustive—in fact urban garden

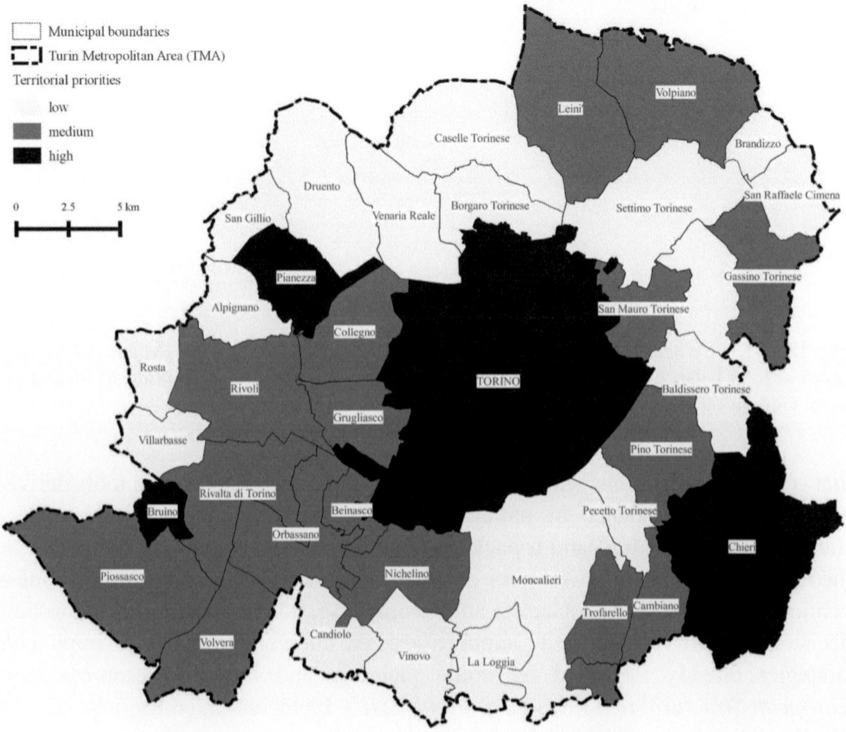

Fig. 15.4 The priorities map of the TMA. *Source* Reworking of the authors from: Gottero (2016)

classification has to be completed and fine-tuned, further on-the-spot surveys need to be carried out and the geo-localisation of farms improved—the proposed method could be replicated in other metropolitan areas, in order to constitute the cognitive bases indispensable for coding, management and planning. As regards the interpretative and evaluation framework, some additional clarification is needed. On the one hand, the set of numerical indicators—valid on the metropolitan scale and aggregated at municipal level, for the assessment and monitoring of the status and change of essential ecosystem services in urban farming provided to humans—lends itself readily to both synchronous and diachronic application, as well as implementation and adaptation to other case studies, especially in the context of spatial planning and strategic environmental assessment instruments. On the other, meta-project tools that combine indications and criteria of spatial targeting—derived from the cognitive framework and map-based tools—with objectives, intervention priorities and specific measures, i.e. permitted interventions and eligible areas, are easily interfaced with the framework of the CAP and RDPs, as well as with that of other European funds. However, these are devices that would require a further phase of analysis and experimentation, as well as fine-tuning of the data and strict control in the field. The existence of other databases, especially on agricultural holdings, as well as the occasional lack of homogeneity and validity of the data used, still remain the most important obstacles, especially in the construction phase of the cognitive framework.

Finally, another question to reflect upon concerns the demand for urban farming. The analysis carried out, supported by telephone interviews, analysis of other databases and a review of the literature, has shown that the demand for urban farming related goods and services, but also simply the number of new requests made by citizens for allotments to be used as family gardens, is extremely high and in most cases superior to the supply or number of public and/or private spaces set aside for this purpose. This proves that most municipalities are not equipped to meet the needs of citizens, and the urban farming carried out in the city is predominantly conventional, with a business model not specifically oriented to the urban market that continues to operate through traditional techniques and markets. And yet, the research carried out on the Turin case study has shown that agriculture in the metropolitan area is very different from that practised in the past. It is neither the agriculture that involves rural areas, nor that more commonly identified as "niche" agriculture. Rather, it is a hybrid, complex form of agriculture that embraces the heritage of the past and the food, social and cultural challenges of the future, an agriculture that rejects the marginalisation of rural areas in order to move closer to the new urban demands.

15.5 Conclusions

The Turin case study has demonstrated how far issues regarding agriculture in metropolitan areas are still nowhere near a solution or any real awareness of their importance, especially for the city. Researchers, experts, decision-makers, policy-makers and citizens are therefore left with numerous responsibilities, especially on the ideological and methodological fronts. The territorial innovation that accompanies these policies and practices, as well as the issues that are still unresolved today, seem to involve two aspects that have hitherto been at the very least underestimated.

The first concerns the problem of territorial identification, i.e. the urban-rural dichotomy widely debated in the literature (see, for example, Mougeot 2000; Russo et al. 2014; Wascher et al. 2015) and at institutional level. This distinction is decisive not only from a conceptual point of view, but also at times operationally, since it is also the cornerstone of the territorial distribution of certain European funds. For example, no measure of the current Italian RDPs provides support for urban agriculture, not least because the definition of "rural territory" remains enigmatic. The identification and eligibility of urban areas is therefore still debatable. What is urban, what is rural? Is there a spatial limit on urban agriculture? Is it still reasonable to consider it marginal, especially from an ecosystem perspective? There are still opposing positions in this regard today. Although to date there is no unambiguous interpretation and definition of a metropolitan area, the analysis conducted on the Turin case study has highlighted the unequivocal presence of a group of municipalities united not only by their spatial contiguity, but above all by relationships of close cooperation (economic, social, cultural, etc.) coinciding with those of the former metropolitan conference, as brought together by shared functional and productive objectives, as well as common physical, morphological and environmental aspects. The most recent definition of *Functional Urban Area* (FUA) promoted by the OECD in 2012 as part of the studies on urban redefinition, supported also by today's *Metropolitan Areas Database*[7] seems to be moving in this direction. Each FUA represents an economic unit characterised by a densely inhabited city core and by an area around which the city's labour market (commuting zone) gravitates. Thus, UA is gradually approaching a metropolitan dimension (Viljoen and Bohn 2014), i.e. it is a phenomenon that involves not only the municipalities of the conurbations, but rather a wider basin, probably more similar to the FUA than to the TMA. The perception is that TMA agriculture is only a part of a more extensive and heterogeneous phenomenon which, by adopting the OECD method, could lead to a metropolitan agriculture, understood as an activity aimed at satisfying urban demand, and to a different spatial classification. Thus defined, the metropolitan agricultural system would seem to represent a complex phenomenon that stands out for its proximity to the urban market and the magnetism that the city exerts on the agricultural and non-agricultural activities of a basin of close urban and extra-urban relations.

Another important aspect that seems to have been neglected so far, especially in Italy, concerns the minimum size of agricultural space and agro-ecosystem services that should be the due of every metropolitan inhabitant. Although to some extent utopian and probably difficult to implement, the proposal to define a minimum amount of rural space, as a concept borrowed from town planning theories—to be found in the literature in Rovai et al. (2013)—namely a minimum amount of agricultural space and services to meet the urban demand of citizens, is becoming increasingly topical and demonstrates the lack of not only quantitative instruments, but also of clear rules and standards for agricultural areas. Is this scenario conceivable? Since it is not possible to act on market dynamics, this aspect could be addressed by working mainly on the dimensioning of new public spaces and on the mechanism of equalisation and compensatory instruments. On the one hand, a minimum amount of agricultural land would have to be defined—which could take the form not only of a social garden, but also of a farmers' market—in relation to the inhabitants who can be settled, and, on the other hand, the negative environmental balance should be eliminated or reduced to a minimum by allocating abandoned or degraded urbanised land, returned to urban farming and given back to farmers for use in off-ground crops, educational and social farms, direct points of sale, energy crops or the functional restoration of soils. However, the feeling is that

[7]In this regard see: http://stats.oecd.org/Index.aspx?Datasetcode=CITIES (accessed: 29/09/2016).

these are instruments that have yet to be codified, an empirical phase and experimental research into new standards and threshold values, not only to keep the expansion of the city in check, but also to restore sustainable food, social and cultural balances.

Acknowledgements This work was supported by ISI Foundation, CRT Foundation and IRES Piemonte of Torino, Italy (Lagrange scholarship, 2015). In addition to Dr. Stefano Aimone (IRES), Dr. Ilaria Vittone (trainee at IRES) and the RDP evaluation staff of IRES, we would like to thank for the methodological support and the data provided: Mario Perosino (Piedmont Regional Authority), Filomena Nardacchione, Michela Sigliano, Lorena Cora and Emilio De Palma (CSI Piemonte).

References

Bhatt, V. C., Kongshaug, R., & McGill University. (2005). *EL 1: Making the edible landscape: A study of urban agriculture in Montreal*. Montreal Minimum Cost Housing Group, School of Architecture, McGill University, Montreal. Accessed on 5 September, 2013 http://www.mcgill.ca/mchg/pastproject/el.

Branduini, P. (2008). *La gestione delle trasformazioni nel paesaggio agricolo periurbano. Permanenze storiche e paesaggi futuri*. In R. C. Venturelli (Ed.), *I paesaggi d'Europa tra storia, arte e natura*. Proceeding of the "Conferenza Trilaterale di Ricerca 2005–2007", Loveno di Menaggio, Como, Italy. Accessed on 20 August, 2015 http://www.urbanagricultureeurope.la.rwthachen.de/files/branduini_2008_paesaggio_agricolo_periurbano.pdf.

Caspersen, O. H., & Olafsson, A. S. (2010). Recreational mapping and planning for enlargement of the green wedges in greater Copenhagen. *Urban Forestry and Urban Greening, 9*(2), 101–112.

Donadieu, P. (2004). *Campagne urbane. Una nuova proposta di paesaggio della città*. Roma: Donzelli.

Gottero, E. (2015). Il paesaggio rurale italiano tra vecchie e nuove politiche agricole. *Territorio, 74,* 134–145.

Gottero, E. (2016). *Agricoltura Metropolitana. Politiche, pratiche e opportunità per l'innovazione territoriale nel torinese* [Metropolitan agriculture. Policies, practices and opportunities for territorial innovation in Turin]. Torino: IRES Piemonte. Accessed on 5 March, 2018 http://www.digibess.it/fedora/repository/openbess:TO082-01941.

Hsin, R. (1996). *Guidelines and principles for sustainable community design. A study of sustainable design and planning strategies in North America from an urban design perspective*. Thesis, Faculty of the School of Architecture at Florida Agricultural and Mechanical University.

Ingersoll, R., Fucci, B., & Sassatelli, M. (2007). *Agricoltura urbana: dagli orti spontanei all'Agricivismo per la riqualificazione del paesaggio periurbano*. Bologna: Regione Emilia Romagna.

Laviscio, R. (2014). *Cultural dimension of urban agriculture*. COST Action Urban Agriculture Europe: Short Term Scientific Mission Report, London. Accessed on 20 August, 2015 http://www.urbanagricultureeurope.la.rwthaachen.de/resource-center/cost-uae-downloads.html.

Lohrberg, F. (2013). *Barcelona declaration on Urban agriculture and the CAP*. Steering group meeting of COST Action Urban Agriculture, June 11–12th 2013, Brussels. Accessed on 20 August, 2015 http://www.urbanagricultureeurope.la.rwthaachende/files/130624_barcelona_declaration_on_urban_agriculture.pdf.

Lohrberg, F., Licka, L., Scazzosi, L., & Timpe, A. (Eds.). (2016). *Urban agriculture Europe*. Berlin: Jovis Verlag.

Mougeot, L. (2000). Urban agriculture: Concept and definition. *UA Magazine*, p. 1.
Piedmont Regional Authority. (2015). *Regional landscape plan*. Accessed on 22 October, 2015 http://www.regione.piemonte.it/territorio/pianifica/ppr.htm.
Rovai, M., Fastelli, L., & Pucci, F. (2013). Verso una pianificazione efficace delle aree agricole periurbane: un nuovo approccio metodologico per la piana di Lucca. In *Proceeding of XXXIV Conferenza Italiana di Scienze Regionali*, Palermo, 2–3 Settembre 2013.
Russo, P., Tomaselli, G., & Pappalardo, G. (2014). Marginal periurban agricultural areas: A support method for landscape planning. *Land Use Policy, 41,* 97–109.
Simon-Rojo, M., Recanses, X., Callau, S., Duzi, B., Eiter, S., Herandez-Jimenez, V., et al. (2016). From urban food gardening to urban farming. In F. Lohrberg, L. Licka, L. Scazzosi, & A. Timpe (Eds.), *Urban agriculture Europe* (pp. 22–28). Berlin: Jovis Verlag.
Tempesta, T. (2011). Paesaggio ed economia. In M. Agnoletti (Ed.), *Paesaggi rurali storici. Per un catalogo nazionale* (pp. 133–148). Roma-Bari: Laterza.
Termorshuizen, J. W., Opdam, A. E. P., & Opdam, P. (2009). Landscape services as a bridge between landscape ecology and sustainable development. *Landscape Ecology, 24,* 1037–1052.
UN—Habitat. (2015). *Habitat III issue papers. 10—Urban-rural linkages*, New York. Accessed on 16 February, 2016 http://unhabitat.org/wp-content/uploads/2015/04/Habitat-III-Issue-Paper-10_Urban-Rural-Linkages-2.0.pdf.
Vidal, R. (2009). *Entre ville et campagne, un paysage à inventer*. Coll. "Ateliers techniques du paysage", Conseil Régional de Bretagne, Rennes.
Viljoen, A., & Bohn, K. (Eds.). (2014). *Second nature urban agriculture: Designing productive cities*. London, New York: Routledge.
Wascher, D. M., Piorr, A., Pintar, M., Kneafsey, M., Sali, G., Corsi, S. L., et al. (2015). FOODMETRES—Metropolitan food planning connecting the local with the global. *Urban Agriculture Magazine*, n. 29, pp. 41–44.
Willemen, L., Verburg, P. H., Hein, L., & van Mensvoort, M. E. F. (2008). Spatial characterization of landscape functions. *Landscape and Urban Planning, 88,* 34–43.
Zasada, I. (2011). Multifunctional peri-urban agriculture. A review of societal demands and the provision of goods and services by farming. *Land Use Policy, 28,* 639–648.

Enrico Gottero (Editor) is lecturer in Landscape and Regional Planning at the Interuniversity Department of Regional and Urban Studies and Planning (Politecnico di Torino), as well as researcher at the Institute of Economic and Social Research of Piedmont (IRES). His main interests focus on the interaction between agriculture and rural-agrarian landscapes, urban-rural linkages, farmland preservation, urban agriculture, agri-food policies and practices. He is also author of several publications on these topics, including at international level.

Claudia Cassatella is an Architect, Ph.D. in Landscape Design/Planning and Associate Professor in Landscape and Regional Planning, Vice-Chair of the Faculty of Planning and Design at Interuniversity Department of Urban and Regional Studies and Planning (Politecnico di Torino). Her research focus on landscape, cultural and natural heritage, at different scale and aimed to different operational areas (landscape plans, strategic plans, management plans of protected areas and UNESCO sites, urban planning, rural development plans and related evaluation tools). She is member of the executive board of UNISCAPE Network of the Universities for the implementation of the European Landscape Convention. She is also author and editor of more than a hundred publications, among which Landscape Indicators (Springer, 2011).

Chapter 16
Conclusion: Ten Good Reasons to Foster an Innovative Vision of Agrarian Urbanism in Europe

Enrico Gottero

This book does not claim to be exhaustive. It is a point of initiation and convergence of several disciplines that share the interests of studies and research centred on planning, management and enhancement of agricultural and rural landscapes. The authors of this book have extensively underscored these mutual relations, showing that a new alliance that addressed the emerging issues of cities and the environment around us is possible. Each paper provides evidence of the need for a multidisciplinary, interdisciplinary and multisectoral approach, and also of an integrated and holistic vision to overcome current weaknesses in open space planning and sectoral programming. Hence, a new field of interaction and comparison must be defined, a common and integrated discussion table to build the technical framework required to synergistically tackle society's issues concerning environmental and urban health, food, poverty and social inequality, quality of life and the well-being of citizens. The concept of agrourbanism should follow this trend. However, is it a viable solution? How can we promote this vision? What are the reasons that should underpin this approach? Why should we support and promote this approach? In the following pages we shall attempt to justify this perspective by discussing some key issues and the ideas behind this vision.

16.1 Contrasting Land Consumption and Reducing Land Degradation

Stopping soil degradation and reducing the pressure of agriculture, energy use and urbanisation are goals established by the most recent international policies on sustainable development (UNCCD 2012). City and countryside should be indis-

E. Gottero (✉)
IRES Piemonte, Via Nizza 18, 10100 Turin, Italy
e-mail: e.gottero@ires.piemonte.it

E. Gottero (ed.), *Agrourbanism*, GeoJournal Library 124,
https://doi.org/10.1007/978-3-319-95576-6_16

solubly linked to contain land consumption, and this can be achieved by green areas and infrastructures that contrast the unhealthy features of cities and the lack of public spaces for leisure time, sports and recreational activities (Donadieu 2005, 2013; Matos and Batista 2013). Hence the need to overcome the concept of solely restraining urban development and sprawl by interrupting the urban continuum, besides considering the countryside as an inanimate space (Matos and Batista 2013). The original conflict between town and country should adopt the synergistic vision of the farmer playing a social role by maintaining territorial and environmental resources, and of the positive externalities for the city (markets and agri-food products, new tourist and recreational economies, services for the community, etc.) (Camagni 1994; Donadieu 2013). This phase of overcoming the concept of the green belt must be necessarily accompanied by gradual awareness of multifunctional agriculture and new geographical proximity relations with urban environments. This must be central to a territorial project designed to fight phenomena, such as urbanisation and agricultural intensification (Mazzocchi et al. 2014; Olivi 2012; Rovai et al. 2013). Ensuring a minimum amount of agricultural space and rural services to guarantee key ecosystem functions provided by agriculture, especially in urban and peri-urban areas, supporting multifunctional and environmental sustainability in agriculture, regulating the real estate market, managing relations with the urban context (land use, distance, environmental restrictions, population density, etc.) will be one of the next challenges faced by agrarian urbanism in the implementation of sustainable development goals.

16.2 Restating the Role of Agriculture in the Urban Environment

The considerable economic regression we notice today is another assumption of urban agriculture. At the same time, urban shrinkage and sprawl in many European cities have given rise to extensive degraded, abandoned or unused spaces. The recent increase in abandoned or unbuilt areas, buildings and industrial districts, abandoned railway and military zones, public green areas or underused and degraded urban parks could offer an opportunity to improve and enhance the quality of urban spaces. These places could be significantly transformed by agricultural practices, such as agricultural parks (see Fanfani's chapter in this book), urban forestation, community gardens, etc. (Haasea et al. 2014; Russo et al. 2014), as evidenced by some significant and heterogeneous experiences in Detroit, Lipsia, Copenhagen, Zurich, Barcelona, etc. (Ingersoll et al. 2007; Ingersoll 2012; Mathey et al. 2015). Despite the possibility of major contamination problems (as a result of deindustrialisation), brownfield regeneration plays a crucial role in promoting healthy and functional urban environments. Consistently with this approach, urban agriculture could be an important tool not only for soil de-sealing and land recycling (BIO 2014; EEA 2016), but also for re-naturalisation and greening of the urban environment. It offers an opportunity to improve social inclusion and to support low income, remediation of polluted sites, maintenance and improvement of ecological corridors, and the production of energy crops.

16.3 Strengthening the Urban-Rural Bond

Despite being scarcely discernible today, the distinction between urban and rural environment has often been attributed to two contrasting functional issues, namely commercial, financial, administrative and institutional significance of the cities, and primary production of food, fibre and energy in rural areas. This dichotomous approach is often also reflected in the dual modern interpretation of population censuses or in European spatial classification systems. However, Santangelo (Chap. 5 of this book) claims that fully understanding the relationship between urban and rural is a challenging feat. Especially in the last decades, globalisation, industrialisation and specialisation of agriculture have deeply changed relations between town and country (Vejre et al. 2016). The bond between urban and rural environments is also the topic of a new institutional debate, such as the *General Assembly of the United Nations* and the most recent *Habitat III Conference*. In this regard, *issue papers on urban and rural linkages* have underscored several core matters that are open to debate and which link the urban framework with the rural one such as, for instance, the expected population growth in urban areas, the subsequent decrease and impoverishment in the rural context, as well as interdependencies concerning access to food, ecosystem and social services, transport, employment and markets. This is a wide range of functions and flows of people, goods, services and capital that is of crucial importance for mankind. Hence, according to these concepts (Dubbeling et al. 2015; UN-Habitat 2015), the maintenance of some agricultural functions, broadly speaking, functions that can ensure a sustainable future for cities and the countryside, and improve environmental quality, especially in the urban fringe, should be a priority in the public agenda of many European countries.

16.4 (Re)-Assigning Value to the Food Supply Chain

Although the reasons to say that food is a basic right of mankind might seem self-evident, relations between food and territory, and reasons that should encourage cities to adopt a targeted policy within the public agenda, are less evident. The issue of food security evokes not only aspects related to the use of food, but also situations that affect availability, as well as economic and physical access to food Dansero et al. (2013); see also Parham (Chap. 8) and Pettenati (Chap. 11 in this book). In many parts of the world, not just in the *Global South*, the availability of land for food production has become a matter of crucial importance, not just in terms of health and food safety (in this perspective see also Diehl et al. in this book). The central nature of these aspects is proven by the proliferation of innumerable institutional documents that attempt to (re)-connect the theme of food with territorial and environmental aspects (Calori 2010) including the latest initiatives, such as the *Seul International Declaration*, *Milan Urban Food Policy Pact*, etc. (Dubbeling et al. 2015). Although urban planners can partly justify this initial lack of interest by arguing that the issue of food is predominantly a "rural" problem, outside the spatial planning agenda, the multifunctional nature of the food system implies profound repercussions on several aspects and fields where planners have a legitimate interest. The growing attention of urban planners for food policies could be traced to what Morgan and Sonnino (2010) call "new food equation", which stems from the latest developments in terms of insecurity (price variation) and food poverty, in climate change and its effects (food supply chain emissions), in the land grabbing phenomenon (especially in Africa and in Asia) (see also Diehl et al. in this book), in the rising incidence of food-related diseases, and in the growing awareness of food scarcity among citizens (Morgan 2009, 2014 and 2015; Vidal and Santini 2014). Food should, therefore, be considered not only in terms of feeding for survival, but also as a set of values that, through the territorial dimension, reveal the historical and cultural aspect of food, and the geography of a place where food is produced (Dansero et al. 2013).

16.5 Reviving Green and Blue Infrastructural Networks

The European Commission (EC) has recently launched several initiatives to promote green and blue infrastructural frameworks, especially as part of EU 2020 biodiversity strategies and within the priorities set out in Europe 2020. Consistently with this trend, the EC has launched a specific strategy, "Green Infrastructure - Enhancing Europe's Natural Capital" (COM/2013/0249), focused on the ecological, economic and social benefits of these "natural remedies" (EC 2013; Rega and Spaziante 2013; Rega in this book). The open network and green space system is also particularly important for the city's metabolism, considering social implications (public health, food education, active citizenship, etc.), and for the health and well-being of citizens (Borgogno et al. 2015; Di Iacovo et al. 2013; La Rosa 2014; Naldi et al. 2015; Zasada et al. 2011). In this process, it is crucial to strengthen and protect the networks of natural and semi-natural areas, such as farmland, in urban and rural contexts, in order to provide diversified ecosystem services, such as biodiversity, carbon sequestration, pollution reduction, microclimate regulation, recreational functions, etc. In this regard, multifunctional agriculture could be a relevant feature of *Green Infrastructure* (GI), especially in terms of management and government of land use and change. Hence, the conservation of these areas depends on maintaining accessibility, an important spatial feature on which the link between users and the integrated system of green spaces is based (Barbosa et al. 2007), and on the conservation and implementation of linear connection and mitigation elements. Thus, some specific agricultural uses such as, for example, urban farms, urban gardens and agricultural parks, could be of considerable importance in either building or strengthening the green and blue infrastructural network (La Greca et al. 2011; Martinico et al. 2013).

16.6 Strengthening Agroecosystem Services and Reducing Disservices

Over the past ten years, the ecosystem services concept—namely a set of functions and benefits (economic, social, environmental, etc.) provided by the ecosystem to mankind—has gradually acquired considerable importance in the context of rural landscape studies (Rega and Spaziante 2013; Scolozzi et al. 2012). Specifically, the agroecosystem has been deliberately simplified by human processes to produce certain goods (Swift et al. 2004). These processes do not only provide but also consume various ecosystem services. Hence, the agroecosystem depends on several supporting and regulatory services that can be defined as biophysical characteristics, if we consider the importance of macro/micro invertebrates and of microorganisms in terms of soil fertility and food production, of insects in terms of nitrogen recycling, pest control and pollination, the quantity and quality of water, as well as the presence of wetlands, riparian vegetation and hedges especially to reduce soil erosion, improve water flow and limit floods. Some conventional agricultural practices can also reduce the service flow, precisely ploughing, cultivation and mechanical harvesting of crops, the use of pesticides and, in a broad sense, agricultural intensification can give rise to the loss of ecosystem services. At the same time, agroecosystems subjected to several ecosystem disservices reduce productivity and increase production costs, such as pests, competition with other plants, etc. For these reasons, the use of agricultural practices and techniques that are compatible with the natural or semi-natural ecosystem network is of the utmost importance to optimise the flow of services provided by and for agriculture. This issue is important both for biodiversity conservation and also to improve productivity, crop yield and farmers' income (Giaimo et al., in this book; Marchetti et al., in this book; Zhang et al. 2007; van Zanten et al. 2014).

16.7 Revitalising Rural Areas and Preventing Abandonment of Farmland

Farmland in the *Global North* is affected by several competing interests and sought-after by several actors (including citizens, regional and local authorities, professional organisations, farmers, trade associations, and environmental organisations), economic sectors, dynamics and the law. This approach presents several consequences, namely a radical change in land use, loss of biological and cultural diversity and identity, a reduction in ecosystem services and functions and, above all, farmland abandonment with all that follows, both in terms of non-agricultural practices and non-agricultural uses, as well as in terms of sustainable agricultural reuse theories (see also Epp et al., and Takatori et al., in this book).

Farmland abandonment has several implications for rural development, bio-economy and social equity (Leung 2003), and involves farmers, especially in terms of production, besides other social groups and stakeholders. Local production capacity and quality, the development of new rural economies (local food chain, rural tourism, etc.), aesthetic values of the landscape and cultural heritage conservation depend on agricultural land conservation. Facing the issue of farmland also means protecting several values of the traditional rural landscape, its artistic, architectural and archaeological heritage, as well as practices and notions of local traditions. However, we still know little about socio-economic and biological causes and effects of land abandonment, especially in the Global North. To date, few studies have explored the causes and effects of farmland abandonment. Some scholars argue that farmland abandonment in Europe is mainly *«due to increasing tax returns on productive lands, conservation policies, and rising importation of agricultural products from other regions»* (Estel et al. 2015, p. 312). In non-European countries, instead, major causes primarily include lack of labour resources and agricultural infrastructures, low product prices and agricultural income, unfavourable land condition, etc. (Tsuchiya and Hagihara 2017). Hence, to prevent farmland abandonment and strengthen a new rural framework based on, for example, youth entrepreneurship, several complementary agricultural activities (environmental education, tourism, welfare, direct sales of agricultural products, social farming, etc.) and environmental services could significantly contribute to create new social relations and innovative local economies (Donadieu 2005; Ferraresi 2013).

16.8 Stating the Social Role of Farmers

In recent decades farmers have gradually acquired a specific social role, not only in terms of food production. During this period agriculture has undergone a radical change, and the farmer has become the main supporter of the production of traditional, healthy and genuine food, as well as the keeper of material memory. Nowadays, farmers play an active role in the management of natural resources and landscape protection. However, the farmer is, first of all, an entrepreneur with a productive purpose, and he mainly follows market rules. Hence, spatial planning can reduce farmland consumption and encourage the reuse of rural buildings but, to date, it has not been able to constrain farmers to comply with the rules of cultivation and farmland use, maintenance of specific characteristics, traditional practices or landscape features, etc. (see also Gottero and Cassatella in this book). Nevertheless, spatial and landscape planning with specific integrations and tools, such as incentives and support mechanisms, tax exemption, quality geo-brands, etc., as well as negotiation and participative decision-making processes, could be an important cognitive and regulatory reference framework. The latest European experiences, such as *Local Development Plans* (LDPs), integrated projects, network contracts, environmental stewardship agreements, *Supply Chain Pact*, etc., have shown how these partnerships can be the most effective tools to respond to the specific demand of a rural landscape (Fanfani 2013; Fanfani in this book).

16.9 (Re)Assigning Historical and Identity Values to Cultural and Traditional Agricultural and Rural Landscapes

In the rural and urban context, agriculture can contribute to the maintenance and transmission of socio-ecological collective memories, preserve open green spaces and increase the resilience of urban population, especially in times of uncertainty and global crisis (Barthel et al. 2015). Agricultural and rural landscapes play an important part in preserving the material memory, traditions and identity of a place, as well as in improving a community's sense of belonging to a specific territory (Cassatella and Seardo 2014; Fanfani 2009; Seardo in this book). Interactions between natural components (climate, morphology, flora, fauna, etc.) and cultural dimensions (ideology, social organisation, technology, etc.) of the rural landscape, which derive from evolutionary processes of the past, manifest not only through the artistic, architectural and archaeological heritage, but also through farmers and their practices, notions and products, besides local environmental resources that are used for this purpose (Agnoletti 2013; Cevasco and Moreno 2013). Thus, the rural landscape is closely bound to the human component and its activity (Antrop 2005; Gambino 2011). However, the recent socio-economic crisis combined with capitalist urbanisation and agro-industrial development (intensification and industrialisation) has expanded the gap and dichotomy between humans and nature. According to Gambino (2011), reconstructing relationships between naturalness and rurality or, as Maffi and Woodley (2010) and Maffi (2010) claim, identifying the deep bond between the work of man and nature, between agriculture and local biodiversity, should be a priority for many European countries, which, at times, mainly focus on an ecological approach.

16.10 Improving Wellbeing and Quality of Life

All decisions concerning maintenance and development of urban areas such as, for example, construction and operation of green areas and cycle paths, initiatives to reduce pollution or restrict vehicular traffic, as well as those affecting public transport, impact quality of life and, in a broad sense, health (Bellaviti 2005). Interaction between regional and urban planning, including health protection (broadly defined), intercept the principle of equity, primarily in terms of service accessibility (cultural, recreational, administrative, social, green, etc.), the concept of participation and involvement of citizens in decision-making, as well as cooperation between different policy areas (Tsourou 2005). The contribution of agriculture to healthier cities is particularly important, considering, for example, the rapid growth and decline of cities and the resulting serious hazard for public health. Moreover, *«city dwellers have the potential to develop city-based policies, which are ecologically sound and which support health»* (Ashton 1992, p. 6). From this standpoint, agriculture plays a crucial role in improving nutrition, quality of life and physical activity. Many cities have exploited such initiatives to launch policies for sustainable use of land and water, improve access to produce and build awareness about public health (Twiss et al. 2003). The need to revive the issue of wellbeing and human health is central to the debate on social justice, mortality and nutritional diseases, which are defined as new inequalities of capitalist society (Morgan 2015). Consistently with this trend, the curative and therapeutic meaning of agriculture, both in the urban and rural context, has gradually drawn the interest of the community not only in terms of benefits offered by the outdoor setting (contact with nature), but also for the positive effects arising from manual labour (Senes and Toccolini 2013). Many studies and research conducted since the '80s have provided evidence of the beneficial effects of natural environments and gardens, besides the positive emotional, cognitive and physical effects (see also Van den Berg et al. 2010). Regarding acute stress symptoms, for example, positive health effects achieved by visiting natural environments, and by practicing horticulture and gardening, are called restorative, and are meant to improve physical, mental and spiritual wellbeing. Some studies have also reported the benefits and impact on quality of life of those who grow their own food. In this regard, Church et al. (2015) say that the increase in urban and peri-urban agriculture results from economic difficulties and the need for fresh food, and above all that people who practice urban gardening are far happier than those who do not. Hence, this approach reveals new social relationships by directly involving citizens to respond to the failings of effective urban public policy, on the basis of civic engagement and participatory democracy. Processes are based on the different functions, including that which Caggiano (2012) calls "natural hunger", as well as a gradual ecological conversion of the city, or according to Uttaro (2012), as genuine re-appropriation of the city to meet community needs centred on health, well-being and, in a broad sense, a more sustainable lifestyle. These are "active citizenship" practices that contribute to the (re)-construction of places, including abandoned and marginal degraded urban areas, and new community places for socialisation (Uttaro 2012).

In conclusion, to reflect on the ways of producing, processing and consuming food, energy and irreproducible resources (land), as well as on public health problems, has become an urgent issue for mankind, especially in metropolitan areas. In order to cope with new lifestyles and emerging consumption patterns in many Western cities, as well as to bridge a disciplinary gap that gives agricultural space the same urban values, a new multidisciplinary approach of reconciliation between city and countryside is needed. It is in this context that the most recent experiences of agricultural planning described in this book are included.

References

Agnoletti, M. (Ed.). (2013). *Italian historical rural landscapes: Cultural values for the environment and rural development*. Dordrecth, Heidelberg, London, New York: Springer Verlag.

Antrop, M. (2005). Why landscapes of the past are important for the future. *Landscape and Urban Planning, 70,* 21–34.

Ashton, J. (Ed.). (1992). *Healthy cities*. Buckingham (UK): Open University Press.

Barbosa, O., Tratalos, J. A., Armsworth, P. R., Davies, R. G., Fuller, R. A., Johnson, P., et al. (2007). Who benefits from access to green space? A case study from Sheffield, UK. *Landscape and Urban Planning, 83,* 187–195.

Barthel, S., Parker, J., & Ernstson, H. (2015). Food and green space in cities: A resilience lens on gardens and urban environmental movements. *Urban Studies*, 1321–1338.

Bellaviti, P. (Ed.). (2005). *Una città in salute: Healthy urban planning a Milano*. Milano: FrancoAngeli.

BIO (by Deloitte). (2014). *Study supporting potential land and soil targets under the 2015 Land Communication* (Report prepared for the European Commission, DG Environment, in collaboration with AMEC, IVM and WU. https://publications.europa.eu/en/publication-detail/-/publication/fdbdf00a-87ac-4c85-8eab-ef60118963c5/. Accessed October 10, 2017.

Borgogno-Mondino, E., Fabietti, G., & Ajmone-Marsan, F. (2015). Soil quality and landscape metrics as driving factors in a multi-criteria GIS procedure for peri-urban land use planning. *Urban Forestry & Urban Greening, 14,* 743–750.

Caggiano, M. (2012). È ritornato il tempo delle ciliegie nei Jardins Partagés di Parigi. *Sociologia Urbana e Rurale, 98,* 28–43.

Calori, A. (2010). Network alimentari locali e politiche urbane. *Agende istituzionali ed esperienze. Territorio, 52,* 57–59.

Camagni, R. (1994). Processi di utilizzazione e difesa dei suoli nelle fasce periurbane: Dal conflitto alla cooperazione fra città e campagna. In F. Boscacci & R. e Camagni (Ed.), *Tra città e campagna. Periurbanizzazione e politiche territoriali* (pp. 13–89). Bologna: Il Mulino.

Cassatella, C., & Seardo, B. M. (2014). In search for multifunctionality: The contribution of scenic landscape assessment. In C. Rega (Ed.), *Landscape planning and rural development. key issues and options towards integration* (pp. 41–60). Cham Heidelberg, New York, Dordrecht, London: Springer.

Cevasco, R., & Moreno, D. (2013). Rural landscape: The historical roots of biodiversity. In M. Agnoletti (Ed.), *Italian historical rural landscapes: Cultural values for the environment and rural development* (pp. 141–152). Dordrecth Heidelberg, London, New York: Springer Verlag.

Church, A., Mitchell, R., Ravenscroft, N., & Stapleton, L. M. (2015). "Growing your own": A multi-level modelling approach to understanding personal food growing trends and motivations in Europe. *Ecological Economics, 110,* 71–80.

Dansero, E., Testa, C., & Toldo, A. (2013). Verso la Smart City partendo dal cibo. In M. Santangelo & A. Vanolo (Eds.), *Smart city: Ibridazioni, innovazioni e inerzie nella città contemporanea* (pp. 135–149). Roma: Carocci Editore.

Di Iacovo, F., Brunori, G., & Innocenti, S. (2013). Le strategie urbane: Il Piano del Cibo. *Agriregionieuropa, 9*(32), 9–16.

Donadieu, P. (2005). Dall'utopia alla realtà delle campagne urbane. *Urbanistica, 128,* 15–20.

Donadieu, P. (2013). L'immaginario dei territori agrourbani o la terra ritrovata. In *Scienze del Territorio* (pp. 59–64). 1/2013, Firenze: Firenze University Press.
Dubbeling, M., Hoekstra, F., Renting, H., Carey, J., & Wiskerke, H. (2015). Editorial: Food on the urban agenda. *Urban Agriculture Magazine*, (29), 3–4.
Estel, S., Kuemmerle, T., Alcántara, C., Levers, C., Prishchepov, A., & Hostert, P. (2015). Mapping farmland abandonment and recultivation across Europe using MODIS NDVI time series. *Remote Sensing Environment, 163,* 312–325.
European Commission (EC) (2013). *Communication from the commission to the European parliament, the council, the European economic and social committee and the committee of the regions. Green infrastructure—Enhancing Europe's natural capital.* (COM/2013/0249). http://eur-lex.europa.eu/legal-content/EN/TXT/?uri=CELEX:52013DC0249. Accessed March 03, 2018.
European Environment Agency (EEA) (2016). *Land recycling in Europe: Approaches to measuring extent and impacts* (EEA report No 31/2016) Luxembourg: Publications Office of the European Union.
Fanfani, D. (2013). Empowerment rurale e rigenerazione paesaggistico-ambientale del territorio agricolo: Una domanda di mobilitazione sociale e 'pianificazione dal basso'. In D. Poli (Ed.), *Agricoltura paesaggistica: visioni, metodi, esperienze* (pp. 179–199). Firenze: Firenze University Press.
Fanfani, D. (Ed.). (2009). *Pianificare tra città e campagna: Scenari, attori e progetti di nouva ruralità per il territorio di Prato.* Firenze: Firenze University Press.
Ferraresi, G. (2013). Neo-rurality: Roots of future in the field. *Scienze del Territorio* (pp. 71–78). Firenze: Firenze University Press. (1/2013).
Gambino, R. (2011). *Il paesaggio tra coesione e competitività*. Trento: Step-Scuola per il governo del territorio e paesaggio di Trento.
Haasea, D., Haasec, A., & Rinkc, D. (2014). Conceptualizing the nexus between urban shrinkage and ecosystem services. *Landscape and Urban Planning, 132,* 159–169.
Ingersoll, R., Fucci, B., & Sassatelli, M. (2007). *Agricoltura urbana: Dagli orti spontanei all'Agricivismo per la riqualificazione del paesaggio periurbano*. Bologna: Regione Emilia Romagna.
Ingersoll, R. (2012). Urban agriculture. *Lotus, 149*(74), 105–117.
La Greca, P., La Rosa, D., Martinico, F., & Privitera, R. (2011). Agricultural and green infrastructures: The role of non-urbanised areas for eco-sustainable planning in a metropolitan region. *Environmental Pollution, 159,* 2193–2202.
La Rosa, D. (2014). Accessibility to greenspaces: GIS based indicators for sustainable planning in a dense urban context. *Ecological Indicators, 42,* 122–134.
Leung, H. L. (2003). *Land use planning made plain* (2nd ed.). Toronto: University of Toronto Press.
Maffi, L. (2010). *Storia di un territorio rurale. Vigne e Vini nell'Oltrepò Pavese*. Milano: FrancoAngeli.
Maffi, L., & Woodley, E. (2010). *Biocultural diversity conservation: A global sourcebook.* Londra: Earthscan.
Martinico, F., Barbarossa, L., La Rosa, D., & Privitera, R. (2013). L'integrazione tra nuove forme di agricoltura urbana e Greenways per la produzione di Ecosystem Services in contesti urbani. *Planum—The Journal of Urbanism, 01,* 1–8.
Mathey, J., Rößler, S., Banse, J., Lehmann, I., & Bräuer, A. (2015). Brownfields as an element of green infrastructure for implementing ecosystem services into urban areas. *Journal of Urban Planning and Development, 141*(3), 1–13.
Matos, R. S., Batista, D. S. (2013). Urban agriculture: The allotment gardens as structures of urban sustainability. In M. Özyavuz (Ed.), *Advances in landscape architecture: InTech* (pp. 457–512). http://www.intechopen.com/books/advances-in-landscape-architecture/urban-agriculture-the-allotment-gardens-as-structures-of-urban-sustainability. Accessed January 20, 2016.
Mazzocchi, C., Sali, G., & Corsi, S. (2014). Fragilità del contesto agricolo periurbano: Uno strumento di analisi per il governo del territorio. *Scienze Regionali, 3,* 51–72.

Morgan, K. (2015). Nourishing the city: The rise of the urban food question in the Global North. *Urban Studies, 52*(8), 1379–1394.

Morgan, K. (2014). The new urban foodscapes: Planning, politics and power. In A. Viljoen & K. Bohn (Eds.), *Second nature urban agriculture: Designing productive cities*. London & New York: Routledge.

Morgan, K. (2009). Feeding the city: The challenge of urban food planning. *International Planning Studies, 4,* 341–348.

Morgan, K., & Sonnino, R. (2010). The urban foodscape: World cities and the new food equation. *Cambridge Journal of Regions, Econ-omy and Society, 3*(2), 209–224.

Naldi, L., Nilsson, P., Westlund, H., & Wixe, S. (2015). What is smart rural development? *Journal of Rural Studies, 40,* 90–101.

Olivi, A. (2012). Oltre il parco e l'orto urbano. Spazio pubblico in movimento e nuovi immaginari urbani. *Sociologia Urbana e Rurale, 98,* 60–72.

Rega, C., & Spaziante, A. (2013). Linking ecosystem services to agri-environmental schemes through SEA: A case study from Northern Italy. *Environmental Impact Assessment Review, 40,* 47–53.

Rovai, M., Fastelli, L., Pucci, F. (2013) Verso una pianificazione efficace delle aree agricole periurbane: Un nuovo approccio metodologico per la piana di Lucca. In *Proceeding of XXXIV Conferenza Italiana di Scienze Regionali*, Palermo, 2–3 Settembre 2013.

Russo, P., Tomaselli, G., & Pappalardo, G. (2014). Marginal periurban agricultural areas: A support method for landscape planning. *Land Use Policy, 41,* 97–109.

Scolozzi, R., Santolini, R., & Morri, E. (2012). Territori sostenibili e resilienti: La prospettiva dei servizi ecosistemici. *Territorio, 60,* 1–9.

Senes, G., & Toccolini, A. (2013). Healing garden: le aree verdi per il benessere dell'uomo. In M. A. Breda & M. C. Zerbi (Eds.), *Rinverdiamo la città: Parchi, orti e giardini* (pp. 65–88). Torino: G. Giappichelli Editore.

Swift, M. J., Izac, A. M. N., & van Noordwijk, M. (2004). Biodiversity and ecosystem services in agricultural landscapes-are we asking the right questions? *Agricultural Ecosystem Environment, 104,* 113–134.

Tsourou, C. (2005). L'Healthy urban planning: Pianificare la città sana. In P. Bellaviti (Ed.), *Una città in salute: Healthy urban planning a Milano* (pp. 47–54). FrancoAngeli: Milano.

Tsuchiya, K., Hagihara, K. (2017). Agricultural landscape: Farmland abandonment and direct payments. In H. Shimizu, C. Takatori, & N. Kawaguchi (Eds.), *Labor forces and landscape management: Japanese case studies* (pp. 83–98). Singapore: Springer.

Twiss, J., Dickinson, J., Duma, S., Kleinman, T., Paulsen, H., & Rilveria, L. (2003). Community gardens: Lessons learned from california healthy cities and communities. *American Journal of Public Health, 93*(9), 1435–1438.

UNCCD (2012). *Zero net land degradation: A sustainable development goal for Rio + 20*. Bonn: UNCCD Secretariat Policy Brief, United Nations Convention to Combat Desertification.

UN—Habitat (2015). *Habitat III Issue papers. 10—Urban-Rural Linkages*. New York, http://unhabitat.org/wp-content/uploads/2015/04/Habitat-III-Issue-Paper-10_Urban-Rural-Linkages-2.0.pdf. Accessed Febuary 16, 2016.

Uttaro, A. (2012). Dove si coltiva la città. Community gardening e riattivazione di spazi urbani. *Sociologia Urbana e Rurale, 98,* 12–27.

Van den Berg, A. E., van Winsum-Westra, M., de Vries, S., & van Dillen, S. (2010). Allotment gardening and health: A comparative survey among allotment gardeners and their neighbors without an allotment. *Environmental Health, 9*(74), 1–12.

van Zanten, B. T., Verburg, P., Espinosa, M., Gomez-y-Paloma, S., Galimberti, G., Kantelhardt, J., et al. (2014). European agricultural landscapes, common agricultural policy and ecosystem services: A review. *Agronomy for Sustainable Development, 34*(2), 309–325.

Vejre, H., Eiter, S., Hernandez-Jimenez, V., Lohrberg, F., Loupa-Ramos, I., Recasens, X., et al. (2016). Can agriculture be Urban? In F. Lohrberg, L. Licka, L. Scazzosi, & A. Timpe (Eds.), *Urban agriculture Europe* (pp. 18–21). Berlin: Jovis Verlag.

Vidal, R., & Santini, C. (2014). Organizzare l'agricoltura delle regioni urbane. *Per una governance alimentare sostenibile. Territorio, 71,* 43–48.
Zasada, I., Fertner, C., Piorr, A., & Nielsen, T. S. (2011). Peri-urbanisation and multifunctional adaptation of agriculture around Copenhagen. *Danish Journal of Geography, 111*(1), 59–72.
Zhang, W., Ricketts, T. H., Kremen, C., Carney, K., & Swinton, S. M. (2007). Ecosystem services and disservices to agriculture. *Ecological Economics, 64,* 253–260.

Enrico Gottero (Editor) is lecturer in Landscape and Regional Planning at the Interuniversity Department of Regional and Urban Studies and Planning (Politecnico di Torino), as well as researcher at the Institute of Economic and Social Research of Piedmont (IRES). His main interests focus on the interaction between agriculture and rural-agrarian landscapes, urban-rural linkages, farmland preservation, urban agriculture, agri-food policies and practices. He is also author of several publications on these topics, including at international level.

MIX
Papier aus verantwortungsvollen Quellen
Paper from responsible sources
FSC® C105338

If you have any concerns about our products,
you can contact us on
ProductSafety@springernature.com

In case Publisher is established outside the EU,
the EU authorized representative is:
Springer Nature Customer Service Center GmbH
Europaplatz 3, 69115 Heidelberg, Germany

Printed by Libri Plureos GmbH
in Hamburg, Germany